Strategic Management in the Wine Tourism Industry

"Strategic Management in the Wine Industry - Competitive Strategies, Wine Tourism Behaviour and New Strategic Tools represents the latest thinking on strategic management from an illustrious group of international scholars that have provided important information for academics, practitioners, and researchers in the wine industry."

—Armand Gilinsky, *Emeritus Professor, Wine Business Institute, Sonoma State University, USA*

"A tour de force in the field of wine tourism research, this book is a rare blend of rigorous academic analysis and actionable insights. The authors' exploration of strategic management within the wine tourism industry is both enlightening and inspiring, making it an essential read for anyone seeking to understand or influence this dynamic sector."

—Gustav Visser, *Full Professor, Department of Geography & Environmental Studies, Stellenbosch University, South Africa*

"In Strategic Management in the Wine Tourism Industry, the authors have masterfully unveiled the complexities of the wine tourism market. Their comprehensive analysis is a beacon of insight, shedding light on the delicate balance between authenticity and profitability. This book is an indispensable resource for scholars and practitioners alike, offering a profound understanding of the strategic intricacies that shape the wine tourism landscape."

—Jolanta Słoniec, *Full Professor, Department of Enterprise Organization, Lublin University of Technology, Poland*

"This seminal work stands at the forefront of academic literature, offering groundbreaking perspectives on wine tourism strategy. The authors skillfully navigate the nuanced interplay of innovation, consumer behavior, and competitive dynamics. Their eloquent discourse is not only a scholarly triumph but also a practical guide that will undoubtedly shape the future of wine tourism management."

—Nikos Georgantzis, *Full Professor, School of Wine & Spirits Business, Burgundy School of Business, France*

Javier Martínez-Falcó ·
Bartolomé Marco-Lajara ·
Eduardo Sánchez-García ·
Luis A. Millán-Tudela
Editors

Strategic Management in the Wine Tourism Industry

Competitive Strategies, Wine Tourism Behaviour and New Strategic Tools

palgrave
macmillan

Editors
Javier Martínez-Falcó
University of Alicante
Alicante, Spain

Bartolomé Marco-Lajara
University of Alicante
Alicante, Spain

Eduardo Sánchez-García
University of Alicante
Alicante, Spain

Luis A. Millán-Tudela
University of Alicante
Alicante, Spain

ISBN 978-3-031-54836-9 ISBN 978-3-031-54837-6 (eBook)
https://doi.org/10.1007/978-3-031-54837-6

This Palgrave Macmillan imprint is published by the registered company Springer Nature Switzerland AG
The registered company address is: Gewerbestrasse 11, 6330 Cham, Switzerland

Paper in this product is recyclable.

Preface

In the esteemed volume *Strategic Management in the Wine Tourism Industry: Competitive Strategies, Wine Tourism Behaviour and New Strategic Tools*, published by Palgrave Macmillan, the authors dissect the multifaceted realm of wine tourism. This academic paper is a testament to the interplay of differentiated interaction of competitive strategy, consumer behavior, and innovative approaches in the wine tourism sector.

The journey into the world of wine tourism begins with an essential quest for clarity in "Toward a Clarification of the Definition of Wine Tourism". This chapter reviews and reflects on existing definitions, proposing a rich understanding that lays a solid foundation for further discussions, underlining the importance of a clear conceptual framework as a prerequisite for advancing wine tourism research and practice, while laying the groundwork for the in-depth explorations that follow. The narrative progresses to the strategic heart of wine tourism. "Cornering Authenticity in Wine Tourism: The Case of Tuscany" delves into the strategic valorization of authenticity, presenting it as a pivotal competitive asset. This exploration is complemented by

"Knowledge Management, Innovation and Business Growth in Wine Tourism: The Case of *Finca La Estacada*" which showcases the transformative power of knowledge and innovation in the sector. The discourse on competitive dynamics is further enriched in "Coopetition Strategies and Wine Tourism Offerings", presenting coopetition as a novel strategy for enriching wine tourism experiences through collaborative competition, emphasizing strategic approaches that can increase the attractiveness and sustainability of the sector.

Turning the focus toward the market, "Message in a Bottle: Generation Y and Wine Purchase Occasions' Perceived Risks and Information Sources" explores the perceived risks and information sources influencing Generation Y's wine purchases, providing insights into the evolving consumer landscape. This is further explored through the lens of gender in "Exploring Emerging Chinese Wine Tourists: Does Gender Matter?" which sheds light on the differentiated preferences of this growing tourist segment. The sector's consumer insights are deepened with "A Systematic Review of Wine Tourism and Customer Behavior Using Bibliometric Analysis", offering a comprehensive bibliometric analysis that maps the scholarly landscape, while "Wine Tourism in Israel: Preferences of Potential Wine Tourists" focuses on the preferences of potential wine tourists, expanding the geographic and cultural understanding of consumer behavior in wine tourism.

Innovative management tools and methodologies are introduced in "Forecasting Wine Tourism: A Survey of Methods", which surveys forecasting methods, highlighting their strategic importance for anticipating sectoral trends. This is synergistically complemented by "Broadening the Idea of Guides and Guiding: A Wine Tourism Roadmap" which reimagines guiding practices in wine tourism, proposing a roadmap for enhancing visitor engagement and experience. Concluding the volume, "Analysing the Experience of Wine Tourists: An Approach from the Wine Routes" synthesizes insights from various wine routes, offering a forward-looking perspective on research and practice. The final chapter serves both as a synthesis of the knowledge acquired throughout the volume and as an insight into future perspectives on the wine tourism sector through the future research lines proposed in the study.

The text thus represents a wake-up call on the fundamental role of strategic knowledge management and innovation in the field of wine tourism, as it explains in detail how the synthesis of organizational knowledge and creativity enhances both the growth of wineries and the wine tourism experience. The treatise goes beyond theoretical reflections and offers pragmatic insights and strategic frameworks that can be useful for professionals in the wine and tourism sectors.

The manuscript contributes to the theoretical foundations of wine tourism and strategic management by clarifying the definition of wine tourism and laying the groundwork for future research, ensuring that academic research is grounded in a common understanding. This theoretical clarity facilitates a more focused and coherent investigation of the complexities of wine tourism. Furthermore, the exploration of competitive strategies, such as authenticity, knowledge management, and coopetition, expands the theoretical boundaries of strategic management in the context of wine tourism, introducing novel frameworks and models that can be applied to other sectors, suggesting wider applicability of the knowledge generated. The manuscript also addresses the theoretical gap in understanding the dynamic interaction between consumer behavior and strategic management, offering novel results for understanding how these elements come together to drive competitive advantage in wine tourism.

For wine tourism professionals, this manuscript offers a set of strategic tools and insights. Discussions on authenticity and coopetition offer practical strategies for differentiating offerings in a saturated market, highlighting the importance of unique and genuine experiences and the strategic value of collaboration with competitors. The focus on consumer behavior, especially as it relates to Generation Y and the new Chinese wine tourists, provides professionals with the knowledge to adapt their marketing and operational strategies to the changing preferences of these key demographic segments. In addition, the introduction of new management tools, such as forecasting methods and innovative targeting practices, presents practical solutions to common industry challenges, from anticipating tourism trends to enhancing visitor experiences.

In essence, this volume is an invaluable scholarly resource that bridges the gap between academic research and practical application in the field of wine tourism. It stands as a beacon of knowledge for academics, industry practitioners, and students, endowing them with a deep, holistic understanding of the strategic intricacies of wine tourism. This book is not merely a collection of research; it is a guidepost for those navigating the dynamic waters of the wine tourism industry.

Here's to a journey of discovery, reflection, and celebration. Cheers!

Alicante, Spain

Javier Martínez-Falcó
Bartolomé Marco-Lajara
Eduardo Sánchez-García
Luis A. Millán-Tudela

Acknowledgments

To Francisca Maestre Pina (1929–2022),

Like the most resilient vines that weather the harshest storms only to yield the finest grapes, she navigated life's challenges with unwavering strength and elegance. Her memory will always be treasured and remembered.

With love and remembrance

Contents

Notes on Contributors

Dr. Claudia Bazzani is Assistant Professor in Rural Economics and Appraisal at the Department of Management, University of Verona. Her research endowers center in the application of experimental methods, such as choice experiments and experimental auctions to investigate consumers' preferences and willingness to pay for food and wine products characteristics. Her research interests also cover wine tourism and consumer perception towards the related services. She has published in leading international journals such as, among others, *American Journal of Agricultural Economics, European Review of Agricultural Economics, Agricultural Economics, Economic Inquiry, Australian Journal of Agricultural and Resource Economics.*

Giovanna Bertella is an Associate Professor at the School of Business and Economics, The Arctic University of Norway. Her research interests are: small-scale tourism, food tourism, rural tourism, nature-based tourism, active tourism, evento management, tourism entrepreneurship, knowledge, and networks. Since 2014 she has been a member of the research program Northern InSights.

Ana Brochado is an economist and professor of management at the ISCTE-Instituto Universitário de Lisboa - Lisbon University Institute and a researcher at DINAMIA'CET-Centre for Socioeconomic and Territorial Studies. She coordinated and taught several courses (bachelor, master, Ph.D., executive education) including econometrics, quantitative methods applied to competition and regulation, consumer behavior, and market research and research methods. She served as the Vice Dean for administrative affairs and Vice Dean for faculty for five years (ISCTE Business School) and coordinated the Ph.D. in Tourism Management. She joined the General Council of ISCTE. She worked for more than one decade as Senior Economist for the Portuguese Competition Authority (AdC) and the Securities Commission (CMVM). She joined the Strategic Board of Compete2020 as an expert in policy evaluation methods. She is an external expert on the European Commission. She has a Bachelor of Science in Economics, a Master of Science in Quantitative Methods, a Ph.D. in Management, and Habilitation in Marketing.

Alessio Cavicchi is a Researcher in Agricultural Economics. He received his Ph.D. in Economics of Food and Environmental Resources from the University of Naples "Parthenope" and a Master of Science (MSc) in Food Economics and Marketing from the University of Reading (UK). His works have been published in several journals, including *Food Quality and Preference*, *British Food Journal*, *Journal of Agricultural Economics*, *Agribusiness: an International Journal*, *International Journal of Wine Business Research and Food Economics*, among others. Previously, he worked in Finance and had a Internship at the European Parliament in Brussels. His main fields of interest in research are consumer food choice, economics of food quality and safety, and innovation in the agro-food sector.

Eli Cohen's research interests are focused on consumers' behavior and perception of the quality of wine, olive oil, and other products. Hospitality and tourism are also of his interest, especially wine tourism, which he studied both in Israel and in Europe. His research has been published in high-ranked journals and presented at professional conferences. He is a member of academic journals and a reviewer of academic journals in wine, tourism, hospitality, food, and more. He is also recognized as an outstanding reviewer of academic manuscripts for academic journals. He taught in international courses in Israel, France, Hong Kong, and Australia.

Dr. James M. Crick is an Associate Professor of Marketing and Entrepreneurship at the University of Leicester (United Kingdom) and an Adjunct Professor of Entrepreneurial Marketing at the University of Ottawa (Canada). His research interests are positioned at the marketing/entrepreneurship interface—sometimes called entrepreneurial marketing. This covers topics, such as internationalization activities, cooperative strategies (e.g., coopetition and value co-creation), dynamic and evolving business models, strategic orientations, and more. Moreover, Dr. Crick is an Associate Editor of the *Journal of Strategic Marketing*, together with being an active editorial board member (and an ad-hoc reviewer) for various outlets.

Prof. David Crick is Professor of International Entrepreneurship and Marketing in the Telfer School of Management, University of Ottawa, Canada. His research interests are positioned at the intersection between international entrepreneurship and marketing. Examples include owner-managers' strategies among start-up firms with limited resources and capabilities alongside an evolving business model; also, internationalization processes and the support requirements of entrepreneurs.

María Ángeles García-Haro is a part-time lecturer at the University of Castilla-La Mancha and holds a Ph.D. in Economics and Business from the same university. Her research interests are co-creation, consumer behavior, corporate image, social media and TICS.

Stella Kladou is an Assistant Professor at the Hellenic Mediterranean University (HMU) in Crete, Greece. She has served as an Assistant Professor at Boğaziçi University in Istanbul, Türkiye and a Senior Lecturer at Sheffield Hallam University in the UK. Before pursuing a PostDoc at HMU, she realized her Ph.D. in cooperation with the Hellenic Open University (Greece), Istanbul Bilgi University (Türkiye) and the University of Rome La Sapienza (Italy), which was partially funded by the Italian Ministry of Foreign Affairs and the Scientific and Technological Research Council of Türkiye. She has published in prominent academic journals, and co-edited books. She has participated and/or led several international and national projects. She is an International Place Brand Specialist for the Place Brand Observer, in the scientific board of international conferences and in the editorial boards of the *Journal of Destination Marketing and Management*, the *Journal of Qualitative Research in Tourism* and the *Journal of Foodservice Business Research*. Currently Stella serves as an Adjunct Faculty Member (Hellenic Open University), and a Visiting Fellow at the Center of Excellence in Food, Tourism and Sports at the American College in Greece. Her research mainly pertains to place branding, cultural and wine tourism, and digital marketing.

Martin Kunc obtained a Ph.D. in Decision Science and is Professor of Management Science at Southampton Business School. He has worked with the Chilean wine industry in innovation, entrepreneur, and tourism. Later, he worked with Champagne region on regional strategic analysis. He also performed market analysis of Japanese sake (rice wine) and Scotch whisky. In terms of tourism, he has published papers on strategic planning and community innovation in journals such as *Tourism Management and Current Issues in Tourism.*

Jorge Linuesa Langreo is Assistant Professor of Business Organization at University of Castilla-La Mancha, Spain. Before returning to academia to earn a Ph.D. in business management, he spent more than ten years in management positions in the building industry. He is interested in issues on behavioural ethics, servant leadership, organizational behavior, and ethical motivations.

Kyuho Lee is Professor of Marketing at Sonoma State University, California, USA. Dr. Lee received his Ph.D. in Hospitality and Tourism Management at Virginia Tech. Dr. Lee also received an AACSB Post-Doctoral Bridge to Business Program Certificate (Marketing) at Virginia Tech in 2009. Dr. Lee has published his works in academic journals such as *Journal of Retailing, Services Marketing, and The Service Industries Journal*.

Ricardo Martínez-Cañas is Associate Professor at the University of Castilla-La Mancha since 2012. He holds a degree in Business Administration and Management from the University of Castilla-La Mancha (1999) and a Ph.D. in Economics and Business Administration from the University of Castilla-La Mancha (2008). He has been visiting professor at the MIOIR—Manchester Institute of Innovation Research of the Manchester Business School (UK, 2009) and at the Catholic University of Eichstaett-Ingolstadt (Germany, 2011). English language skills (B2 certified by EOI) and basic knowledge of German.

Javier Martínez-Falcó, Ph.D. is Assistant Professor in the Management Department of the University of Alicante (Spain) and Research Fellow at Stellenbosch University (South Africa). In the field of research, he focuses his interest in the wine business, specifically in the areas of wine tourism and wineries' intellectual capital, on which he has written hundreds of publications in the form of articles, book chapters, and contributions to conferences. He has also participated in several national and international research projects, being lecturer of strategic management in the economics and business degrees of the University of Alicante.

Bartolomé Marco-Lajara is Professor of Business Organization at the University of Alicante (Spain). He holds a Ph.D. in Economics, within his doctoral dissertation focusing on strategic alliances management. His primary research interests encompass strategic management and tourism management. He has authored several books, book chapters, and international articles related to teaching methodology and the aforementioned areas. He is Member of the Tourism Research Institute at the University of Alicante since its foundation and has participated in many publicly and privately funded projects, often as Lead Researcher. He served as Assistant Dean of the Economics Faculty at the University of Alicante for eight years and currently holds the position of Head of the Department of Management at the same university.

Luis A. Millán-Tudela is Assistant Professor in the Department of Management at University of Alicante and holds Ph.D. in Business, Economy and Society with an MBA and a bachelor's degree in Mechanical Engineering. His research centers on business longevity, particularly focusing on the factors influencing the survival of companies. This research is conducted within the "Dirección Estratégica, conocimiento e innovación en una economía global (DECI-GLOBAL)" research group. He also teaches management-related subjects in different undergraduate programs, as well as in an MBA.

Mobina Mohammadi is a Ph.D. student in the Economics and Business programme at the University of Castilla-La Mancha. His research interests are related to ethical leadership, servant leadership, social capital, wine tourism and social entrepreneurship.

Juan Luis Nicolau is the J. Willard and Alice S. Marriott Professor of Revenue Management in the Howard Feiertag Department of Hospitality and Tourism Management in the Virginia Tech Pamplin College of Business. He has published in *Strategic Management Journal, Omega, Marketing Letters, Journal of Business Research, Economics Letters, European Journal of Operational Research, Journal of Travel Research, Annals of Tourism Research,* Tourism Management, or *International Journal of Hospitality Management.*

Dr. Jacques-Olivier Pesme is the Director of the Wine Research Centre at the University of British Columbia (Canada). Working for several wine organizations in America, Europe and Oceania, he is highly involved in wine regional development issues through world-class action research and tailor-made programs designed for wine industry leaders. Before joining UBC, he was previously Professor in Wine Business at the University of Adelaide (Australia) and Professor and associate Dean at Kedge Business School in Bordeaux (France) where he founded the Wine & Spirits Academy and was a founding member of the Institute of Vine and Wine Sciences (ISVV).

Prof. Helena Rodrigues holds a Ph.D. in Management-Specialization in Marketing at ISCTE-IUL, Lisbon, Portugal. She received her B.Sc. and M.Sc. degrees in Marketing from ISLA-Lisboa and Laureate. She has worked for global Asset Management Banks and in the Marketing Department for a Global Pharmaceutical Company.She lectures Marketing at ISCTE-IUL and Tourism Marketing at Universidade Católica Portuguesa. Her main research is in Medical Tourism,

Wellness and Health Tourism, Service Quality, social media, and Higher Education.

Pablo Ruiz-Palomino is Full Professor of Business Management at University of Castilla-La Mancha, Spain. He earned a Ph.D. in Business Administration at this same University in november 2008, which was awarded the 1st Prize in the 2010 (IV) Edition of FORETICA-MSD for Business Ethics and Corporate Social Responsibility Research (International Call). His research interests are ethical leadership, organizational behaviour, and social capital in business, tourism, and hospitality contexts.

Rafael Morales Sánchez is a Associate Professor of Business Organization at University of Pablo de Olavide, Spain. His research interests are related to business ethics, corporate social responsibility, sustainability, competency-based management.

Eduardo Sánchez-García, Ph.D. is Assistant Professor in the Department of Management at the University of Alicante (Spain). His research is framed in the field of strategic business management, focusing on firm innovation, new technologies, circular economy, supply chain management, value creation, sustainability, and entrepreneurial orientation in the wine industry. He has published several articles in JCR indexed journals, books, and book chapters and has participated in numerous international scientific congresses.

Cristina Santini, Ph.D. is Assistant Professor at Università San Raffaele in Rome, Italy. Her research interests are: participatory research approaches; food and wine business; sustainability; ecopreneurship; innovative teaching methods and Business Strategy.

Donna Sears, Ph.D. is an Associate Professor of Marketing in the F.C. Manning School of Business at Acadia University. When she moved to Nova Scotia wine country, her background in studying hedonic consumption experiences led her to establish a research program in wine business with a focus on wine marketing and tourism. A long-standing research project with a local wine tour provider inspired her to be curious about the role of guides and guiding in wine tourism. Her involvement in Tourism Nova Scotia and applied research have allowed her to observe the growth of the grape and wine sector firsthand.

Ricardo Sellers-Rubio is a marketing Professor at the University of Alicante (Spain), where he is the Head of the marketing department. His research interests are focused on the tourism industry, wine marketing, distribution channels, and marketing performance measurement. He has authored many articles in referred journals, conference proceedings and edited books. He has published in such journals as the *European Journal of Operational Research, Omega*, *European Journal of Marketing*, *Journal of Business Research*, *Journal of Travel Research*, *International Journal of Hospitality Management, Tourism Economics*, *Industrial Marketing Management*, and *International Journal of Service Industry Management*, among others.

Seunghun Shin is an Assistant Professor in the School of Hotel and Tourism Management in the Hong Kong Polytechnic University. His current research interests include social media analytics, tourists' spatial interaction behavior and metaverse marketing in hospitality and tourism. His research has been published in *Tourism Management*, *Journal of Travel Research*, *International Journal of Hospitality Management*, *International Journal of Contemporary Hospitality Management*, *Journal of Hospitality Marketing & Management*, *Journal of Travel & Tourism Marketing*, *Tourism Review*, among others.

Dr. Katia Laura Sidali is Associate Professor of Rural Economics and Appraisal at the Department of Management, University of Verona. In 2009, she received a double doctoral degree in International Cooperation and Policy for Sustainable Development issued by Almamater Studiorum University of Bologna and a "Ph.D. in Agricultural Sciences" from Georg-August University in Goettingen (Germany). Her scientific interests are related to sustainable consumption of beverages and foods, wine tourism agri-food certifications and agritourism.

Sara Vaz Mendes holds a master's degree in marketing from ISCTE Business School, her main research interests include Wine consumption in younger generations.

Ahmet Usakli is an Associate Professor in the Department of Tourism Administration at Boğaziçi University, Türkiye. He earned his doctorate from Gazi University (Türkiye) and obtained a master's degree from the University of Nevada Las Vegas through the US Department of State's Fulbright Scholarship Program. His research interests revolve around consumer behavior in hospitality and tourism, destination marketing and management, as well as statistical and methodological aspects of hospitality and tourism research.

Terrance G. Weatherbee, Ph.D. is a Professor of Management, teaching and studying at the F.C. Manning School of Business at Acadia University. He tries to understand how historical understanding by academics and practitioners shapes knowledge mobilization and organizational action. His latest research program is focused on how wine business actors in an emerging region strategically operationalize the past and history to establish the legitimacy of the sector. Quite often he finds himself researching other things when his colleague sidetracks him into various loosely related areas of wine tourism. Fortune has him living in the heart of Nova Scotian wine country.

List of Figures

List of Tables

Part I

Introduction

1

Toward a Clarification of the Definition of Wine Tourism

Claudia Bazzani, Jacques-Olivier Pesme, and Katia Laura Sidali

1.1 Introduction

Since 2016, the World Tourism Organization (UNWTO) Conference has highlighted the importance of wine tourism to the socio-economic development of rural territories. For several destinations, winemaking is essential to their history and identity and has become a critical element in the region's brand image. Hence, wine tourism represents an opportunity for many areas worldwide to diversify and boost tourism, promote local

C. Bazzani · K. L. Sidali
Department of Management, University of Verona, Verona, Italy
e-mail: claudia.bazzani@univr.it

K. L. Sidali
e-mail: katialaura.sidali@univr.it

J.-O. Pesme (✉)
The University of British Columbia—Wine Research Centre, Vancouver, BC, Canada
e-mail: jo.pesme@origins.wine

J. Martínez-Falcó et al. (eds.), *Strategic Management in the Wine Tourism Industry*,
https://doi.org/10.1007/978-3-031-54837-6_1

economic development, involve different professional sectors, and bring new resources to the primary industry.

Even if the modern history of wine tourism is relatively recent, it is a rapidly growing sector. A report[1] from the International Centre for Spirits, based on various national sources and statistics, estimates that 70 million wine visitors were in the top 12 most visited countries in 2018. According to the study, the USA, Italy, and France led the way with 30, 14, and 10 million visitors annually. These numbers, albeit estimations, show how vital wine-related tourism activities have become in the central wine-producing countries.

Consequently, the surge in visitor interest in wine has prompted numerous regions to embrace the concept of wine tourism. Alongside winery visits and acclaimed wine trails, often considered the primary attractions (Dodd, 1995; Mitchell & Hall, 2004), wine destinations have introduced additional visitor-oriented services to enhance the overall customer experience. For countries and regions, wine tourism holds an added allure as it is a competitive distribution channel for direct-to-consumer (DTC) sales of wine and associated products (Getz & Brown, 2006). By delivering a unique customer experience on-site, wine tourism contributes to establishing brand loyalty, deepens knowledge of wine regions, fosters a sense of belonging, and cultivates stronger consumer relationships (Bruwer & Alant, 2009). Furthermore, wine tourism plays a pivotal role in the economic development of wine regions (Molina et al., 2015). It is a complementary link between the wine production and tourism sectors (Hojman & Hunter-Jones, 2012).

The combination of positive microeconomic impacts on wineries and the impressive national attractivity figures explain why governmental bodies often take pride in announcing growing numbers of wine visitors, sometimes at the expense of providing precise information. However, the interpretation of these figures can vary due to the ambiguity surrounding the definition of wine tourism activities. The Committee on Tourism and Competitiveness (CTC) of the UN World Tourism Organization (WTO), the global parent organization for tourism, defines Gastronomy

---

[1] Refer to International Center for Spirits, *le classement des pays champions de l'oenotourisme* by J. L. Laboissière.

Tourism as a "type of tourism activity which is characterized by the visitor's experience linked with food and related products and activities while traveling. (…) Wine Tourism or Eno-tourism, a subtype of Gastronomy Tourism, refers to tourism whose purpose is visiting vineyards, wineries, tasting, consuming and/or purchasing wine, often at or near the source".

From an academic perspective, research on wine tourism emerged during the 1990s (Carlsen, 2004; Getz, 2000) and primarily focused on three main topics: destination strategies, wine producers' tourism operations, and wine tourist behaviors (Faugère et al., 2013). Surprisingly, there needs to be more emphasis on clarifying the definition of wine tourism, although several authors have attempted to outline its scope. The definitions provided by Hall (1996), Hall et al. (1997, 2009), and Getz (2000) are frequently cited as a reference. Other aspects have been added to this definition (Marzo-Navarro & Pedraja-Iglesias, 2010), such as local attractions, experiencing the lifestyle of the local people, or staying overnight at vineyards.

Defining the scope of wine tourism is fundamental when assessing and comparing the sector's significance. Understanding what constitutes wine tourism and what does not is essential and strategic. In addition to the existing confusion surrounding the notion, it has been observed that almost every wine region in the world has its understanding of wine tourism activities and has subsequently enacted regulations or practices pertaining to them. For instance, in Europe, one of the final outputs of the project "VINTUR" has been the delineation of a charter of the European Wine Tourism to establish "guidelines to meet the requirements of tourists and consumers" (European Union, 2005). However, the proposal of standard wine tourism practices can still be challenging due to the different legislative regulations of the activities among the European countries (Alonso et al., 2015).

This contribution seeks to establish the foundation for a universally accepted and applicable definition of wine tourism. The diversity in understanding the concept of wine tourism, as underscored in the research, necessitates additional efforts for any initiative aiming at a clear definition. The primary goal of this contribution is to offer an overview of how wine tourism is defined in major global wine regions. It also seeks to question the relevance of figures or data that report successes in wine tourism policies. Additionally, it highlights the difficulty of comparing figures across countries due to significant regulatory divergences. Ultimately, the contribution encourages stakeholders to contemplate the imperative need for a clear, precise, and consensus-based definition of wine tourism.

In this regard, this contribution echoes the work of Phillip et al. (2010), who emphasized agritourism and how understanding any phenomenon requires a basic understanding of its fundamental defining characteristics. Specifically, the authors pointed out that, despite the frequent use of this terminology to "evoke this form of tourism, it had never been clearly defined until then" (Phillip et al., 2010).

Consequently, collaboration has been established with the Great Wine Capitals network, a global network representing renowned wine regions worldwide to explore the reality of wine tourism in major wine-producing regions. An exploratory approach has been adopted, primarily based on analyzing documents such as reports and regulations to define the concept of wine tourism and its related activities (Bowen, 2009).

1.2 Materials and Methods

The Great Wine Capitals and Selected Regions

The Great Wine Capitals (GWC) is a network of major global cities in both the northern and southern hemispheres, which share a key economic and cultural asset to encourage travel, education, and business exchanges between prestigious wine regions. The network was founded in 1999 to achieve excellence in tourism, business services, and education within renowned wine regions. The peculiarity of the GWC Network is that it is the only existing international network dedicated to wine tourism encompassing "Old" and "New" wine regions, combining the experience of stakeholders, academics, and representatives of public bodies (Great Wine Capitals, 2023).

To incorporate a diverse range of wine regions encompassing various cultural, territorial, and economic conditions, we have selected representative samples from both the "New" and the "Old" worlds.

Data Gathering

Contacts were initiated with members of wine organizations within the GWC network to ensure comprehensive data gathering. These contacts were approached and requested to provide official or academic references, governmental texts, guidelines, laws, or any other pertinent material that could define wine tourism within their respective regions or countries. The information sought after was twofold: (1) a general definition of wine tourism, taking as a reference the ones proposed by Hall et al. (1997) and Getz (2000), and (2) a list of the activities that can legitimately be associated with wine tourism.

Accordingly, as an attempt to use a representative sample of wine areas, which may differ in terms of cultural, territory, and economic conditions, we focus on regions both from the "New" and "Old" worlds. Regarding the "New World", we selected case studies from both hemispheres and three different continents: Mendoza (Argentina), San Francisco—Napa Valley (USA), and Adelaide—South Australia (Australia), while for the "Old World", Verona (Italy), Bordeaux (France), and Bilbao—Rioja (Spain) were taken into consideration.

Table 1.1 comprehensively summarizes the collected documents utilized in the present study. Its structure is as follows: the first column identifies the study area; the second column presents the names of the organizations within the GWC network that were contacted and provided the documentation. Moving on to the third column, we outline the sources, i.e., the specific documents that were gathered for the purpose of analyzing a general definition of "wine tourism" in each country. Finally, the last column details the recommended documentation provided by the contacted organizations, which served as the primary sources for identifying the activities associated with wine tourism.

Table 1.1 Selected documents for each region/country

Country	The organization providing the documentation	Sources used for the general definition of "Wine tourism"	Sources used for the definition of activities associated with wine tourism
Mendoza—Argentina	Fundación ProMendoza	"Observatorio Economico del Turismo del Vino de la Republica de l'Argentina", adapted from "World Tourism Organization"	Bozzani, M. L. (2012). "Turismo Enologico en Argentina" Universidad Nacional de Mar del Plata, Facultad de Ciencias Economicas y Sociales, Retrieved from Nulan, on January 2023. ACOVI, Asociación de Cooperativas Vitivinícolas Argentinas, (2018). "Impacto del Enoturismo 2016"
Adelaide—Australia	Department of Primary Industries and Regions, Government of South Australia (PIRSA)	Hall, C. M., Cambourne, B., Macionis, N., & Johnson, G. (1997). Wine tourism and network development in Australia and New Zealand: Review, establishment and prospects. *International Journal of Wine Marketing*, *9*(2), 5–31	Australian Grape & Wine (2021). "2021–22 Pre-Budget Submission" Government of South Australia (2018). "South Australia International Wine Tourism Strategy"
Bordeaux—France	Conseil National de l'œnoturisme	Not explicitly defined	Assises nationales de l'œnoturisme (2018). Label "Vignobles & Découvertes" Guidelines & Audit

(continued)

Table 1.1 (continued)

Country	The organization providing the documentation	Sources used for the general definition of "Wine tourism"	Sources used for the definition of activities associated with wine tourism
Verona—Italy	Ministry of Agricultural, Food and Forestry Policies	Ministerial decree n.2779, 12/03/2019. "Linee guida e indirizzi in merito ai requisiti e agli standard minimi di qualità per l'esercizio dell'attività enoturistica"	Ministerial decree n.2779, 12/03/2019. "Linee guida e indirizzi in merito ai requisiti e agli standard minimi di qualità per l'esercizio dell'attività enoturistica"
Napa—USA	Visit Napa	Napa County "Winery Definition Ordinance", n°947	Napa Valley Tourism Improvement District (2015). "Management District Plan 2015–2025"
Rioja—Spain	Cámara Oficial de Comercio, Industria y Servicios de La Rioja	Rioja wine, Academy Diploma in Rioja Wine Tourism (2018)	Rioja wine, Academy (2018). Diploma in Rioja Wine Tourism DOCa Rioja Control Board (2021). "Rioja Monitor de enoturismo de la Denominacion de Origen Calificata Rioja"

Table 1.1 sheds light on the fact that, in most cases, diverse sources of information were referenced to provide the general definition of "wine tourism" and its associated activities. Notably, region-specific documents were identified for Australia, Spain, and the USA, whereas national-level documentation was collected for the other selected countries. The situation in Italy deserves special attention. For years, the wine tourism sector has been implicitly equated with agritourism and, as such, has been subject to a rich legislative body. In particular, Italian laws 730/1984 and 96/2006 provide that the agricultural company's tourism activities are connected to agriculture/viticulture, which remains the company's fundamental role. With the Ministerial Decree n. 2779 of 2019, wine

tourism in Italy is no longer implicitly equated with agritourism but explicitly. Specifically, the Italian legislator recognizes the experiential nature of wine and amplifies the number of activities that can be invoiced with the agricultural entrepreneur's tax benefits. In the case of Napa Valley, an ordinance defines "tours and tastings" within wineries but does not explicitly outline the specific associated activities. On the other hand, France does not offer an "official" definition of wine tourism. However, the country does provide a quality label for a wine tourism destination. Established in 2009, the "Vignobles & Découvertes" label is awarded for a period of three years by Atout France, based on recommendations from the Conseil Supérieur de l'Oenotourisme (Superior Council of Wine Tourism). This label is bestowed upon tourist and wine destinations that offer a range of diverse and complementary products, including accommodations, restaurants, winery tours and tastings, museums, events, and more. Its purpose is to facilitate the organization of visitors' stays and guide them toward qualified services. The "Vignobles & Découvertes" label represents a collective brand applicable to a destination and not a single winery, and its management and awarding procedure are regulated by a set of rules. In France still, it is worth noting a report from Odit-France (Observation, Development, and Tourism Engineering) enumerating the services offered in the wine sector to attract tourists without, however, delineating a precise scope (Bouzdine-Chameeva, 2011). Finally, in the case of Argentina, the World Tourism Organization proposed a general definition of wine tourism. Regarding the description of the associated activities, documentation (i.e., a bachelor's degree thesis) from the National University of Mar del Plata and ACOVI (Asociación de Cooperativas Vitivinícolas Argentinas) was suggested.

Data Analysis

A text analysis has been implemented to analyze the content of the gathered documentation (Kuckartz, 2013). Specifically, two distinct approaches have been adopted for the analysis of the general definition of wine tourism and for the analysis of the associated activities. In both cases, the data from each region have been assembled together. However, for the general definition of wine tourism, a quantitative approach has been implemented to depict the occurrence of the words in the text. Given the objective of drawing a consensual picture of wine tourism, this kind of analysis allowed us to identify the terms that are most frequently associated with the wine tourism concept. We implemented the "Wordcloud" package for Python (Oesper et al., 2011), which, based on the frequency of the words present in a text, generates a "cloud" of words with different sizes. The higher the dimension of the word, the higher the number of times the word is present in the text. On the other hand, to analyze wine tourism activities, we used a more standard, systematic qualitative data analysis with the inductive identification of thematic categories and the development of a category system (Kuckartz, 2019).

1.3 Results

One noteworthy initial finding of this research revolves around the challenges encountered while collecting official information regarding wine tourism and its associated activities. Given the nature and mission of the organizations comprising the GWC network, we initially anticipated that such inquiries would elicit straightforward and readily available responses. However, it became apparent that obtaining prompt answers or receiving comments such as "we don't have one at the regional level" or "we will explore if something exists at the national level" indicated a certain difficulty in addressing our seemingly simple question.

Interestingly, our straightforward query was consistently met with various obstacles and complexities. This observation highlights the inherent complexity surrounding the availability and accessibility of comprehensive information regarding wine tourism and its related facets.

Analysis of "Wine Tourism" General Definition

Going into detail with the description of our analysis, Fig. 1.1 is the result of the "WordCloud" analysis concerning the definition of wine tourism.

As expected, the words "wine" and "tourism" are the most frequently cited. These are followed by the word "tasting". Similar emphasis is given to the terms related to the wineries (e.g., "wineries", "business", "winery") and the vineyards. Interestingly, the words such as "tours" and "visits" have similar sizes to the word "activities", suggesting the connection between these terms. In the cluster of activities, we can include the ones related to the interactive, recreational aspect of wine tourism, which are described by Byrd et al. (2016) as ancillary activities. In this context, these activities can be resembled by the terms "didactic" and "cultural". Although these words signaling the ancillary activities share less emphasis than the core (wine) activities, they are still clearly visible in the text cloud. What follows in size/importance are the words related to the territory, namely "area" and "place". However, interestingly, the small size of the word "region", which should recall the concept of designation of origin or terroir, indicates that this term has been scarcely used. Finally, limited visibility is held by the items related to gastronomy,

Fig. 1.1 Text analysis of "Wine Tourism" general definition

such as "food" or "gastronomic". Interestingly, these terms have smaller dimensions than the words that can be connected to the core product (i.e., wine), such as "grape" or "production".

Analysis of Activities Associated with Wine Tourism

As the second step in our analysis, we examined the documentation related to the eligible activities. Through an inductive approach, we were able to summarize the emerging patterns from our data and determine the main trends associated with wine tourism (Thomas, 2006). Specifically, out of the collected information, four categories have been identified:

1. *Tasting and visit to vineyards*: As expected, "tasting" is revealed to be the most relevant activity that characterizes each region under consideration, followed by the tours to the cellars, vineyards, and wine routes. The didactic and educational function of the tours is generally enhanced. In the case of Italy, harvesting is mentioned as an activity connected with wine tourism and that the winery's owners can offer, either to their guests or in the case of didactic farms. On the other hand, the case of Australia is also of particular interest since touring the vineyards is regularly forbidden for hygienic reasons and thus may not be considered a possible wine tourism activity.
2. *Merchandising and communication activities*: The commercialization of wine and all the associated marketing activities are prominently featured in the selected documentation. These activities encompass a wide range of practices, including purchasing products directly from wineries, advertising, online promotions, and the utilization of business cards. Interestingly, in most instances, promotional efforts extend beyond the winery itself and encompass the entire region, particularly in the case of designated origin areas in France and Italy. This approach serves as a destination marketing strategy, recognizing the importance of promoting the region alongside the individual wineries.

Notably, both documentation provided for France and Napa explicitly acknowledge that wine tourism activities can significantly contribute to the regional economy. Furthermore, the integration of digital tools creates an intersection between promotional activities and communication/information activities. This convergence extends to analog tools such as signage for routes and wineries, the use of informative materials, and the utilization of tourism information centers.

3. *Recreational activities*: These cover concerts, cultural events, exhibitions, and activities to entertain children during their parents' visits (this was mentioned only in the case of Rioja), both at the regional level and at the level of the single winery.
4. *Gastronomy*: The gastronomic aspect has been regarded as a distinct category within the context of wine tourism. However, it is crucial to note that considerable heterogeneity exists among countries regarding the definition and significance attributed to the "role" of food in wine tourism. In the case of Argentina, the connection with the local "gastronomy" is often mentioned, but no more details are given regarding the food service. In France, the criteria for obtaining the "Vignobles & Découvertes" label include restaurants offering regional cuisine and a wine list predominantly composed of local wines. However, there is no specific requirement concerning whether a winery must have its restaurant or not. In Italy, it is explicitly specified that wineries can only sell and serve cold food products that have been prepared by the winery owners. Moreover, it is recommended that these products originate from the same area where the winery is located, emphasizing a regional connection. Finally, the Italian legislator suggests the selling of food and agricultural products labeled with the PDO (protected designation of origin) and PGI (protected geographical indications) certifications. Similarly to the food aspect, in a limited number of countries, we have come across references to wine bars and restaurants (Australia, Spain, and Argentina).

Finally, it is important to point out that in all the selected documentation, the relevance of respecting safety and hygienic standards, as well as safeguarding the agricultural land and local architecture, is frequently mentioned. Similarly, some operative recommendations are generally provided, which cover opening hours, the seasonality of the wine tourism activities, and the use of trained personnel for welcoming and interacting with the tourists.

1.4 Discussion and Conclusions

The Challenge of Defining Wine Tourism

As the study demonstrates, formulating a universally accepted definition for wine tourism at the international level is fraught with complexity and challenges (Charters & Ali-Knight, 2002). In light of the examples discussed in this paper, here are some reasons why accepting a clear wine tourism definition is a challenging issue.

One of the primary complexities in defining wine tourism internationally lies in the diverse experiences wine regions offer worldwide. Each region boasts unique characteristics, including winemaking practices, cultural heritage, terroir, and local traditions (Mitchell et al., 2009). For instance, in France, wine tourism often emphasizes the historical and cultural aspects of winemaking, including wine festivals (Houghton, 2008), while in Australia, it may focus more on the adventurous exploration of vast vineyard landscapes when authorized or an intimate wine cellar experience at the winery. These contrasting approaches make it challenging to establish a standardized definition encompassing the myriad of global wine tourism experiences. We have noticed that in Latin-speaking countries, people use the term "enotourism", and it can encompass not only wine-related activities but also some broader cultural experiences (Lignon-Darmaillac, 2014), while the English term wine tourism focuses primarily on wine-centric activities (Byrd et al., 2016). Such linguistic variations make harmonizing definitions across languages and cultures a complex task.

The question of national and/or regional regulations is undoubtedly one of the most understandable limits to a harmonized consensus about wine tourism. Each country has its own set of regulations governing wine production and tourism, but surprisingly, this is less the case for governing wine tourism. Italy is a unique example where comprehensive national legislation supports wine tourism. One thing that agritourism and wine tourism have in common is the particular activities or events that can occur on a farm/winery. Although generally speaking, all types of events are usually accepted, only those strictly connected to agricultural and wine-growing activities can benefit from the tax relief derived from the status of a farming entrepreneur. Unfortunately, the decision on the merits of the connection principle is left to the regions, and therefore, there may be bureaucratic delays in the practical application. France provides a certification label. Napa, with its Winery Definition Ordonnance, offers a regulatory framework, however without defining precisely what wine tourism is.

The case of food mentioned in the text analysis illustrates well this disparity of different regulations existing across wine regions. In Rioja, South Australia, and Argentina, going to wine bars or restaurants is one of the activities wine tourists can experience, while in Italy or Napa restaurant activities are not associated with wine tourism.

Last but surely not least, wine regions rely on tourism as a vital source of revenue, contributing to local economies and employment opportunities (López-Guzmán et al., 2011). As such, countries may have differing interests in defining wine tourism to suit their specific economic goals. Some nations prioritize attracting international visitors to boost tourism, while others focus on domestic tourism promotion.

Why Define Wine Tourism?

Agreeing on an international definition of wine tourism is an intricate endeavor due to the multitude of wine tourism experiences, cultural and linguistic differences, regulatory variations, economic considerations, and regional factors. While efforts are pursued to establish guidelines and certifications within countries, achieving a consensus on a global scale remains challenging. Nevertheless, agreeing on an applicable definition for wine tourism in every wine region would be relevant and wise for many simple reasons.

First and foremost, a universally accepted definition would bestow consistency and lucidity upon consumers, businesses, and regulatory bodies. It would ensure that tourists, irrespective of their geographical or cultural background, possess a crystal-clear comprehension of what to anticipate when venturing into a wine region. This uniformity would elevate the overall experience for visitors, empowering them to make well-informed decisions regarding their travel plans.

Despite the efforts and accomplishments attained by the GWC network over the past two decades, a standardized definition would also facilitate the advancement and promotion of wine tourism as a global industry. A shared understanding of wine-related activities and comparable metrics would enable wine regions to collaborate more effectively, discuss best practices, and collectively market themselves to a broader audience. Consequently, this ensures that visitors are shielded from being misled or disenchanted by biased or exclusive claims regarding a region's wine offerings or tourism activities. Like the hospitality sector, establishing guidelines for quality, safety, and authenticity would bolster consumer trust and confidence in wine tourism destinations.

Such a definition would assist governments and regulatory bodies in developing policies and regulations tailored to the needs of the wine tourism sector. It would permit the reporting of consistent figures about the sector. It would provide a framework for addressing issues such as licensing, zoning, and environmental sustainability and potentially quantify seriously the number of wine tourists. Therefore, an accepted clarification would facilitate the implementation of effective regulations supporting the development of a sustainable and responsible wine tourism industry.

As previously mentioned, it is essential to emphasize the significance of a universal definition for wine tourism in stimulating international recognition and fostering networking opportunities for wine regions. Such a definition would empower them to actively engage in forums and conferences, promoting dialogue and collaboration with other wine-producing regions. This exchange of ideas and experiences would encourage cross-cultural understanding and drive the expansion of wine tourism on a global scale.

To summarize, reaching a consensus on a comprehensive definition of wine tourism across all wine regions is relevant and prudent. It offers the benefits of consistency, clarity, and consumer protection. Moreover, it supports the development of markets, the formulation of policies, and the sharing of knowledge within the wine tourism industry. Furthermore, it enhances international recognition, networking opportunities, and overall growth in wine tourism worldwide.

The research suggests that the definition of wine tourism should recall the vineyards of the area, the cellars, and wineries as business units, focusing the activities on tasting and tours with an educational scope. Regarding the definition of the activities, an international consensus is challenging to find due to specific regional regulations and economic considerations.

This exploratory study can stimulate future research endeavors and ignite a constructive discussion regarding the definition of wine tourism. However, it is essential to acknowledge the limitations presented within this study. Primarily, the sample size was constrained, warranting the need for future research to incorporate a broader range of documentation from additional wine regions and countries. Additionally, we strongly advocate for a more comprehensive analysis that entails interviews and group discussions with other experts in the field of wine tourism, including wine producers, to capture their perspectives on the challenges associated with this rapidly evolving domain.

Considering the global tourism sector's continuous growth, alongside the wine industry's anticipated development, it is highly probable that wine tourism will persist in its progressive trajectory. Given its relatively recent emergence and subsequent expansion during the 1990s, the imperative for clarification becomes increasingly evident, particularly as wine tourism finds itself in a stage of rapid development. Consequently, the time has undeniably arrived to establish a consensus-based definition, for presently, when addressing wine tourism, regions resemble players engaging in the same game yet adhering to different sets of rules.

References

Alonso, A. D., Bressan, A., O'Shea, M., & Krajsic, V. (2015). Perceived benefits and challenges to wine tourism involvement: An international perspective. *International Journal of Tourism Research, 17*(1), 66–81.

Bouzdine-Chameeva, T. (2011). L'oenotourisme: portrait du secteur en 2010 des vignobles bordelais. In J.-F. Trinquecoste (Éd.), *Le vin et ses marchés.* Féret.

Bowen, G. A. (2009). Document analysis as a qualitative research method. *Qualitative Research Journal, 9*(2), 27–40.

Bruwer, J., & Alant, K. (2009). The hedonic nature of wine tourism consumption: An experiential view. *International Journal of Wine Business Research, 21*(3), 235–257.

Byrd, E. T., Canziani, B., Hsieh, Y. C. J., Debbage, K., & Sonmez, S. (2016). Wine tourism: Motivating visitors through core and supplementary services. *Tourism Management, 52*, 19–29.

Carlsen, P. J. (2004). A review of global wine tourism research. *Journal of Wine Research, 15*(1), 5–13.

Charters, S., & Ali-Knight, J. (2002). Who is the wine tourist? *Tourism Management, 23*(3), 311–319.

Dodd, T. H. (1995). Opportunities and pitfalls of tourism in a developing wine industry. *International Journal of Wine Marketing, 7*(1), 5–16.

European Union. (2005). *Vintur, projects and documents.* Retrieved from: https://keep.eu/projects/1090/VINTUR-ESPACE-EUROP-EN-DE-L–EN/ on April 2023.

Faugère, C., Bouzdine-Chameeva, T., Pesme, J. O., & Durrieu, F. (2013, January 16). *The impact of tourism strategies and regional factors on wine tourism performance: Bordeaux vs. Mendoza, Mainz, Florence, Porto and Cape Town.* http://doi.org/10.2139/ssrn.2201563

Getz, D. (2000). *Explore wine tourism: Management, development & destinations.* Cognizant Communication Corporation, New York.

Getz, D., & Brown, G. (2006). Critical success factors for wine tourism regions: A demand analysis. *Tourism Management, 27*(1), 146–158.

Great Wine Capitals. (2023). *The Great Wine Capitals global network: A world of excellence.* Retrieved from: https://www.greatwinecapitals.com/about/, on April 2023.

Hall, C. M. (1996). Wine tourism in New Zealand. In *Proceedings of tourism down under II: A tourism research conference* (pp. 109–119). University of Otago.

Hall, C. M., Cambourne, B., Macionis, N., & Johnson, G. (1997). Wine tourism and network development in Australia and New Zealand: Review, establishment and prospects. *International Journal of Wine Marketing, 9*(2), 5–31.

Hall, C. M., Johnson, G., Cambourne, B., Macionis, N., Mitchell, R., & Sharples, L. (2009). Wine tourism: An introduction. In *Wine tourism around the world* (pp. 1–23). Routledge.

Hojman, D. E., & Hunter-Jones, P. (2012). Wine tourism: Chilean wine regions and routes. *Journal of Business Research, 65*(1), 13–21.

Houghton, M. (2008). Classifying wine festival customers: Comparing an inductive typology with Hall's wine tourist classification. *International Journal of Culture, Tourism and Hospitality Research, 2*(1), 67–76.

Kuckartz, U. (2013). Qualitative text analysis: A guide to methods, practice and using software. *Qualitative Text Analysis*, 1–192.

Kuckartz, U. (2019). Qualitative text analysis: A systematic approach. In *Compendium for early career researchers in mathematics education* (pp. 181–197). Springer International Publishing.

Lignon-Darmaillac, S. (2014). L'œnotourisme, redécouverte des valeurs patrimoniales des vignobles historiques, développement des vignobles du nouveau-monde. *Cultur (Revista de Cultura e Turismo), 8*(3).

López-Guzmán, T., Rodríguez-García, J., Sánchez-Cañizares, S., & José Luján-García, M. (2011). The development of wine tourism in Spain. *International Journal of Wine Business Research, 23*(4), 374–386.

Marzo-Navarro, M., & Pedraja-Iglesias, M. (2010). Are there different profiles of wine tourists? An initial approach. *International Journal of Wine Business Research, 22*(4), 349–361.

Mitchell, R., & Hall, C. M. (2004). The post-visit consumer behaviour of New Zealand winery visitors. *Journal of Wine Research, 15*(1), 39–49.

Mitchell, R., Hall, C. M., & McIntosh, A. (2009). Wine tourism and consumer behaviour. In *Wine tourism around the world* (pp. 115–135). Routledge.

Molina, A., Gómez, M., González-Díaz, B., & Esteban, Á. (2015). Market segmentation in wine tourism: Strategies for wineries and destinations in Spain. *Journal of Wine Research, 26*(3), 192–224.

Oesper, L., Merico, D., Isserlin, R., & Bader, G. D. (2011). WordCloud: A Cytoscape plugin to create a visual semantic summary of networks. *Source Code for Biology and Medicine, 6*(1), 1–4.

Phillip, S., Hunter, C., & Blackstock, K. (2010). A typology for defining agritourism. *Tourism Management, 31*(6), 754–758.

Roberts, C. W. (2000). A conceptual framework for quantitative text analysis. *Quality and Quantity, 34*, 259–274.

Thomas, D. R. (2006). A general inductive approach for analyzing qualitative evaluation data. *American Journal of Evaluation, 27*(2), 237–246.

Part II

Competitive Strategies in the Wine Tourism Sector

2

Cornering Authenticity in Wine Tourism: The Case of Tuscany

Cristina Santini, Alessio Cavicchi, and Giovanna Bertella

2.1 Introduction

There are many reasons why companies operating in the wine business decide to improve their portfolio activities and start running wine tourism. Companies could work on direct sales to boost their profits or create new ways to contact customers and reinforce their loyalty.

C. Santini (✉)
Università San Raffaele, Rome, Italy
e-mail: santini.cristina@gmail.com; cristina.santini@uniroma5.it

A. Cavicchi
University of Pisa, Pisa, Italy
e-mail: alessio.cavicchi@unipi.it

G. Bertella
UiT The Arctic University of Norway, Tromsø, Norway
e-mail: giovanna.bertella@uit.no

C. Santini
Laboratorio Sismondi per gli Studi Rurali, Pisa, Italy

J. Martínez-Falcó et al. (eds.), *Strategic Management in the Wine Tourism Industry*, https://doi.org/10.1007/978-3-031-54837-6_2

Other wineries follow the wave and do what others do, hoping this will create good opportunities for them. Some organisations are interested in contributing to local development and see wine tourism as an effective tool for strengthening local economic systems.

The chance to benefit from the tourism business has attracted many companies who have decided to invest resources. However, wine tourism represents a challenging activity that requires time, resources and competencies. The balance between costs and benefits is hard to manage, and wineries must also deal with visitors' expectations when they plan their wine tourism experience. Many wineries have shaped their tourism offer, underlining the aspect of authenticity; this could represent an asset and a starting point for planning the development of a wine tourism business.

This chapter explores the challenges that companies must face in wine tourism, focusing on the Tuscan wine tourism system and then highlighting issues that can be relevant and valid for other regions and areas. The selected case provides inputs that can be generalised since the business is composed mainly of small companies that make authenticity critical assets when defining their offer. Thus, global dynamics are affecting the emerging competitive scenario: companies must reflect upon the drivers of profitability in the wine tourism business. We will adopt a narrative approach focusing on two main issues: on one side, the resources that companies can offer for promoting wine tourism and attracting visitors, and on the other, the dynamics that affect the market (and wine tourists) and the competitive structure of the industry.

The chapter will start by describing the business of wine tourism in Italy and Tuscany; we will then outline the role of authenticity when shaping wineries' strategy. Then, we will change the perspective and focus on the organisational issues that characterise wine tourism. Later, we will outline the dynamics that affect the emerging competitive scenario and some criticalities that can influence the profitability of companies investing in wine tourism.

2.2 Wine Tourism in Italy and Tuscany

In Italy, wine tourism represents a very relevant business.

An impressive number of wineries offer tourism experiences that add something to the traditional wine tour and tasting package by promoting something with deep roots in the local heritage and culture. Examples include visiting wine museums or exhibitions dedicated to wine or participating in local festivals. Under this perspective, wine tourism is a world rich in possibilities: for example, Italy has 56 wine museums, and 22 are embedded only in Tuscany and Piedmont.

It is difficult to find precise data about wine tourism in Italy since there is no specific official census addressed to this sector, and there are multiple data sources; one of the most relevant in the field is The Italian Food and Wine Tourism Report.

Wine tourism in Italy is a vast and relevant phenomenon that covers 58% of the Italian tourist population. According to the data, the number of wine tourists in Italy in 2023 accounted for 9.6 million. The analysis of the activities pursued by food and wine tourists shows that the interest covers culinary heritage in general, events linked to food and wine and themed itineraries. New needs have emerged, including increased attention to sustainable issues.

According to the latest report released in April 2023 by Nomisma, the typical wine tourism actor is a small family winery. This typology accounts for 39% of the wineries offering wine tourism experiences and has found a broad diffusion in the south of Italy. The 14% of wineries attract tourists with their historical or relevant buildings because of their architecture. The 12% of wine tourism offers belong to wineries with a solid and famous historical brand name. The 11% is represented by wineries settled in places where the landscape plays a leading role.

Most wineries can host people within their agritourism, B&B or hotels. Besides wine tours, services include healthy food, cooking classes, sports and recreation (especially where a pool is available) and cultural events. There is a high presence of women in wine hospitality (73%). Moreover, companies need help reaching tourists since 44% of wineries are far from traditional tourist routes. Moreover, they need help reorganising their openings during weekends and festivities due to a lack of professionals involved in the business. Lack of competencies is a critical issue, and according to Nomisma, companies in general (1 out of 4) experience difficulty finding employees with experience in the field.

Tuscany is one of the leading destinations. Given the presence of cities that attract impressive tourism flows, like Florence, it is clear that the population of actual and potential wine tourists is impressive. Furthermore, the concept of "made in Tuscany" wine improves the product's perceived value; some regions in Tuscany have a remarkable global reputation in wine, such as the Chianti or the Montalcino areas.

In Tuscany, the mix of regional wine reputation, the presence of some strong wine company brands or regional product brands, like Brunello or Chianti, and a considerable flow of potential wine tourists have boosted companies' interest towards wine tourism, following the rising interests of tourists.

Therefore, many companies have invested time and resources in developing wine tourism.

Tuscany has about 12,700 wineries—according to the data released by the Tuscan Regional Government—that cover more than 60,000 hectares of vines. The average surface of a winery in Tuscany is about 4.7 Hectares, which means that most of the companies are relatively small or micro, and the number of large companies is minimal. There are 15 cooperative cellars. Most of the wineries are concentrated in the province of Florence (18.44%), Arezzo (20.5%), Siena (19.7%) and Grosseto (19.5%). The size of the companies varies and depends on location: in the province of Florence—which includes the Chianti Classico Area—the average size of wineries is more significant than in other provinces. The phenomenon of merging and acquisition in the local wine industry has improved the size of wineries in some areas, while in other areas of Tuscany, wineries remain small.

2.3 The Emerging Scenario

The growing popularity of wine tourism and the increased number of services and activities planned and realised by wineries open new scenarios for wineries.

Given the growing population of wineries that open their facilities to visitors, companies have felt pressure to improve the quantity and quality of offered services.

Some wineries have chosen to address their offer to some selected market segments: we can cite, for example, the wineries that have decided to focus on cycle wine tourism (Morpeth, 2009). Differentiation can be a successful strategy if the local market becomes too crowded.

A sort of circular dynamic takes place in wine tourism: the growing popularity of wine tourism offered wineries an answer for improving their profits; thus, wineries have expanded their activities by offering wine tourism services; as a consequence, the number of wineries operating in the industry improved, and companies started feeling the pressures deriving from increasing competition. So, wineries had to define their positioning in the wine tourism offer, considering they must also address their service to foreign visitors. Companies must define their strategy and choose what to offer, how to position themselves and how to compete against local competitors. Following market requirements could be a strategy, as well as differentiating wineries' offers by focusing on a specific target or offering a unique experience. Sometimes, developing an offer that focuses on authenticity could be a viable solution, as we will see later.

As wine gained popularity among tourists, new standards for the wineries have emerged: many categories of visitors expect a particular typology of services or a high degree of professionalism.

The various combinations of wine tourism packages have created a differentiation and an organisation of services and offers ranging from essential solutions to more complex and articulated ones.

Wineries whose services and activities can be considered basic could have some difficulties in facing the competition of companies with a broader offer and many options for tourists. Nevertheless, the literature underlines that wine tourism demand differs according to the type of visitor: the studies aiming at profiling wine tourists (see, for example, Bruwer et al., 2018; Bruwer & Lesschaeve, 2012) show the emerging differences across different categories of wine tourists, according to not only to their traits but also on the degree of involvement in wine.

Some tourists have developed a need for more services, and the "level" of wine tourism offered is higher than the basic one. Some cases justify the availability of an essential wine offer. It can happen when the winery is in an area new to the wine tourism or when the company deliberately choose to keep its services simple and basic coherently with the winery image. In the first case, when the winery settles in a territory with limited development in wine tourism, the rural dimension of the activity prevails on the need for professionalism and augmented services. Furthermore, where the wine business is at its embryo phase of development, entrepreneurs must face a lack of skills and competencies for developing a professional wine tourism offer. The described scenario opens a renewed interest in authenticity in the wine tourism business.

2.4 Authenticity

Being authentic represents a strategic asset for some wineries. Authenticity is a helpful tool for promoting wine tourism; wineries often include it in their messages. We searched the internet using a combination of keywords (authentic/authenticity, Tuscan, Wine tours, wine, tourism) and then selected a few results appearing on Google. The aim is to provide examples of how companies perceive authenticity and employ it to design their tourist offer.

Lapi and Tour promotes wine tours underlining the authenticity of their offer: "It is a full immersion in the bucolic atmosphere of authentic Tuscany". The proposed activities are "a walk through the vineyards and olive groves of our estate during which you will visit the small Church of San Giorgio, which has been peacefully observing our vineyards and olive trees since 1109, to then go to the cellar (…). At the end of the walk, in the shade of the white mulberry tree, on the suggestive panoramic terrace, you can taste 3 glasses of wine born from the fruits of the San Giorgio vineyards, authentic Tuscan products: Platter of cold cuts, cheeses and honey, bruschetta and our EVO oil" (from the website: https://www.tuscanywineries.it/en/san-giorgio-a-lapi/).

In this case, authenticity is represented by the bucolic atmosphere and the simple activities organised by the company, consisting of a walk, a tour of the cellar, a visit to the facilities (the church) and wine-tasting with some pairing food.

The winery Solatio links instead the concept of authenticity to cultural heritage: "For three generations, we have been producing organic wines and extra virgin olive oil, and we select wines and niche products from the authentic Tuscan tradition" (from the website: https://www.i-tuscany.com/en-gb/).

We can see how authenticity is linked to visitors' experience: "An authentic Tuscan experience, so we love to define the tastings of food, wine and extra virgin olive oils that we can offer at our A Solatìo farmhouse, starting from the simplest, the intermediate and the most exclusive dinner to be held in the cellar surrounded by hundreds of wines or, in summer, on the panoramic terrace in the middle of our vineyards: not to be missed!" (Solatio).

Also, in this case, the activities pursued are simple and focus on tasting.

Cantine Angeli combines authenticity with family: "FAMILY. AUTHENTICITY. HOSPITALITY. Our family owns a Tuscan wine cellar, where we have been selecting wine for three generations. We have a love for our territory and environmentally friendly production methods in order to offer our customers the highest quality".

The Diacceroni winery underlines the linkage with local tradition, making an experience genuinely authentic: "An expression of true Tuscan tradition. Immersed in the vibrant heart of Tuscany, discover the secrets of wine and its culture. A journey for all of your senses, the extraordinary experience of taste and flavour awaits you in the picturesque atmosphere of the Tuscan countryside. Walk through the vineyards with us and taste the wines of our organic farm. You can also enhance the taste experience by combining it with a lunch or dinner that includes typical Tuscan products" (diacceroni, https://www.diacceroni.com/en/portfolio-item/wine-tasting/).

Torre a Cona refers to authenticity as a model, the model of the Tuscan winery, that shapes visitors' ideals and expectations: "AN AUTHENTIC TUSCAN WINE ESTATE IN FLORENCE HILLS Torre a Cona Wine Estate is one of Tuscany's best-kept secrets. Nestled in the Florence hills, with unforgettable views over the sweeping countryside, the striking 18th-century villa makes an immediate impression. (…). Visit the vineyards and historic cellars to delve into the estate's history before tasting our produce in the poignantly restored barn, now the wine shop. As a final flourish to an authentic Tuscan experience, the Rossi di Montelera family has opened their home to holidaymakers seeking a relaxing gourmet stay in an idyllic setting".

There is a broad debate about authenticity in the literature: it is a complex issue that brings together different aspects of the topic. The topic is also examined by using different approaches that focus on various issues that range from the perception that people have about an experience to the set of relationships with local and cultural heritage (Kim & Bonn, 2016).

Tourism seeking authenticity also distinguishes visitors from the mass (Overend, 2012). Personal interpretations can define what is authentic; subjectivism has a heavy component, and the experience is analysed according to the perception and feelings of the visitor, who can assess something authentic although it is not entirely authentic (Jamal & Hill, 2004). Authenticity has two components, feeling and knowledge: something can become authentic because it is linked to the local heritage and traditions and is rooted in the past. Under this perspective, keeping things as they were in the past can guarantee authenticity. The examples mentioned above provide a comprehensive overview of authenticity applied to the wine industry in Tuscany, and they all underline the linkage between the company and the local culture and atmosphere for shaping an authentic experience in wine tourism.

As we have previously seen, besides the presence of a segment of companies with strong financial capabilities and competencies that can afford the development of an offer that can provide services for the high-end of the market, there is a considerable amount of small wineries that have decided to invest time and money to develop the wine tourism business. Therefore, for those who face difficulties when entering wine tourism, there is an emerging need to develop a positioning in wine tourism that is aligned with their resources and capabilities. Therefore, developing an offer that reduces positioning that focuses on authenticity—as seen from the example—could be a successful choice. Nevertheless, defining what is truly authentic is hugely challenging, as we have previously underlined: it can be authentic what sticks to tradition and what reflects what has been done for years, or what is perceived as authentic, so it is linked to the idea that visitors have about a specific place or experience and how they should be.

Furthermore, more than authenticity is needed to guarantee that visitors will accept a basic wine tourism offer. It is a risky and not proper use of the concept of authenticity. The real issue is that the wine tourism business has set some standards that wineries must respect.

2.5 The Wine Tourism Experience

Undoubtedly, tourists who visit a winery have an experience through the set of activities that the winery plans. For this reason, we have emphasised in the chapter the role of experience and its strategic implications for wine tourism and wineries.

The concept of experience has several facets and has been explored using many approaches and disciplines (Santos et al., 2021). The concept of experience has been vastly explored in tourism: in general, tourism activities are planned to engage visitors. In wine tourism, there is a strong interaction between tourists and some elements that scholars define as "winescapes" (Hall & Mitchell, 2000; Nowak & Newton, 2006). Winescapes consist of a combination of natural settings, heritage, facilities that belong to the winery, the other complementary products and services, signage and people at the winery. The concept of winescape gives an idea of the complex system of variables that belong to the wine tourism system.

Therefore, wine tourism encapsulates several experiences: tasting, purchasing and also attending events that are linked to local cultural heritage or gastronomic offers (Getz & Brown, 2006; Hall & Mitchell, 2000). Nevertheless, it is undoubtful that the multifaceted characteristics of the wine tourism experience are promoted by the characteristics of the wine product itself, with its hedonic traits (Charters, 2006). In other words, there is a deep interdependence between product (wine) characteristics and the complexity of the experience.

Visitors who choose a wine tourism destination have several motivations: They can be attracted by a place (Orth et al., 2012), a product or a winery. In some cases, tourists want to add something different to the tourism experience they are already having. In this scenario, visitors book a tour near where they are staying for their journey with the aim—most of the time—to do something different from the local core attraction.

Thus, as Plasait (2007) outlines, visitors choose a destination because of the attractiveness and availability of services and experiences.

Whatever their motivation is, experience covers a leading role in wine tourism. The literature that explores experience in wine tourism is extensive. We can say that the experience is multidimensional (Santos et al., 2019). It covers personal, relational and environmental aspects (de Freitas Coelho et al., 2018). It is bent over time since it starts with the pre-consumption, up to a post-consumption phase, where the visitor recalls what has happened (et al., 2002). Therefore, visitors have some expectations related to the multidimensional dimension of their experience: they may want to feel something new, perceive some emotions or seek interaction with other people, for example. Expectations shape what visitors search for when booking a wine tour or selecting a place to visit. Experience has a subjective shadow that is hard to interpret if the service provider has little information about the visitor.

There is a link between the visitor's personality traits and his/her choice of destination. This is because, for example, the personality of the tourist shapes the perception of what scholars call the "personality" of the destination.

Therefore, tourists with certain personality traits would seek specific characteristics of the tourism experience (Kovačić et al., 2022).

What people expect from a tourism experience is related to who they are and their traits. The features of an experience of tourism, and wine tourism, differ, and there are several (see among the others, Mehmetoglu, 2007); teach offer of tourism experience has differences in the impact on tourists and their behaviour. For example, the degree of willingness to share the experience changes according to the type of experience offered (Su et al., 2021), not only to the personality of tourists.

Companies should reflect on the role that experience has in their tourism business.

First, they can manage the experience that wine tourists have by offering services according to customers' needs and personalities. Wineries can usefully employ a system of consumer profiling.

Visitors' attributes shape the perception they might have of the offering experience. The degree of wine knowledge can arouse the expectation people have for wine tours.

Visitors that have a more profound knowledge—or wine lovers, as some scholars refer to it—of wine want to gather as much info as they can from their wine tours because they want to taste and buy wine from the cellar from their visits (see, among others, Bruwer, 2003).

Scholars have shown that a deep understanding of visitors' profiles can help wineries to design a successful strategy for their wine tourism offer and plan the activities (Priilaid et al., 2020). Therefore, some scholars have outlined how difficult it can be to design a proper set of activities when there are so many competitors in the surrounding areas: Priilaid et al. (2020), make a clear referral to the principles that inspire the blue ocean strategy designed to carve out a market space where the degree of competition is low when planning wine tourism activities. This approach is wholly motivated by the growing competition in the sector and the expectations that wine tourists might have. Wineries can employ touristic experience as strategic leverage: in wine tourism, experience is a valuable means for strengthening customers' loyalty (Nowak & Newton, 2006). Therefore, improving the knowledge of information gained from consumers leads companies to know better what visitors need. Some researchers have explored the degree of visitors' satisfaction when they pursue a wine tour (see, for example, Barbierato et al., 2021). It becomes important to understand how visitors formulate their expectations, how they approach a wine tourism experience and to see how companies successfully manage the information they gather to plan their strategy and their offer.

2.6 The Activities

Wine tourism is multifaceted (Haller et al., 2020): it is made up of different activities that the visitor can pursue and has various aims. As some scholars outline, it represents a meaningful opportunity for companies to improve their profits, a leverage for strengthening customer relationships (Getz & Brown, 2006) and improving a rural area's overall competitiveness and attractiveness (Carlsen, 2004). A company that invests in wine tourism can choose among different types of activities to offer. Background research has explored the organisation of activities in the wine tourism business (see among the others Santini, 2019) and the implications for the supply chain of wineries and organisation management.

Generally, all planned activities can be organised following the flow of phases that characterises the touristic experience: pre-visit stage, visit and post-visit stage.

Some authors (Guedes & Rebelo, 2019) talk about activities off-premises and on-premises, highlighting how some phases can be outsourced in a supply chain approach.

In the pre-visit stage, companies can offer different services for booking wine tourism activity and help visitors arrange their journey.

Tourists might need help booking and collecting information about the company and its tourist offer. The winery can decide to provide a booking service or to outsource this activity to a service provider; according to the decision, the company needs to invest resources and manage the activities. Information is another key issue; even in this case, the company can decide to perform the required activities or outsource them. Informing the potential visitor requires a combination of tools and duties, according to the degree of detailed information that the winery wants to provide to customers that could include: information about the area, the production process, the characteristics of the product, the characteristics of the winery and its tourism offer. Location helps in reducing the amount of information that the company can deliver. If the company is settled in a well-known area that customers know, it becomes easier to communicate to people the characteristics of a place. In the above-described scenario, the company is facilitated.

Furthermore, the presence of associations (wine routes, wine producers, etc.) or institutions actively involved in promoting local wines by informing tourists can reduce the amount of resources that wineries should dedicate to communication and informing activities.

The winery can decide what tools can employ to inform potential visitors: it can go through the company website or hire agencies to reach the aim.

During the visit, the company has to arrange several activities that can take place in the cellar, vineyards and spaces dedicated to welcoming tourists. Several activities (such as weddings, parties or festivals) can be organised in the vineyard, in the cellar or in the spaces designed to host visitors. The activities include harvesting, biking, picnic and audio tours.

According to Guedes & Rebelo (2019), some off-premise activities that can happen outside the winery facilities can be organised to enrich the tourist experience, like boat or train tours or walking, golf or fishing activities, among others.

The range of activities that can be organised within winery facilities varies from the classic visit to the winery to tasting and other related occupations, such as art exhibitions, concerts or other events that can be hosted inside the companies' facilities. Depending on its facilities winery's offer can include hosting, lodging and dining services.

The more complex the activities are, the more wineries are asked regarding resources and competencies. Over the years, the number of activities that can be undertaken in a wine tour has increased, and wineries have started running new services and events. Offering activities in some areas of the winery (for example cellar) represents a way to spread the fixed cost represented by that asset. After the visit, wineries can offer different services to reinforce the visitor's linkage and improve their loyalty by recalling their visiting experience. In this scenario, we can outline the experience of wineries and wine clubs that can be a valid tool for improving sales (Santini & Faraoni, 2010). Then other tools that belong to the sphere of customer relationship management can be helpful for the company in collecting information to improve and better the service.

The organisation of all the activities can represent, on one side, the chance to improve tourist offer attractions, on the other, a cost that can be quantified in terms of financial resources and time spent on organising and running all the activities.

2.7 Industry Organisation and Institutional Pressure

Investing in wine tourism can also create some difficulties for companies.

Motivations behind difficulties can vary. In some cases, wineries need more competencies or resources to face a new business; other difficulties, such as the changes in the competitive scenario, can represent a severe challenge for companies.

Consider the impact of unexpected events on tourism: the COVID-19 experience has taught wineries that managing the unexpected requires enormous efforts and is stressful.

Also, the increasing degree of industry competition can hurt companies: they must develop adaptable strategies. In the case of the wine business, the growing population of companies offering similar services can threaten a company's competitive positioning.

As an activity, wine tourism defines *duties* that a company must accomplish, providing a set of stakeholders and a system of rules. Wine producers already have their system of relationships and rules to observe. Since they deal with producing wine—an agricultural product—they have established relationships with stakeholders that can widely differ from those in the tourism industry. Scholars, when they deal with the above-described scenario, employ the term "cluster", which somehow underlines not only the numerosity and the range of stakeholders but also the system of relationships they have.

There is a partial intersection between the wine and the tourism cluster, and this can create some difficulties for wineries. As some authors outline (Dressler & Paunovíc, 2021), the fact that these two clusters only partially "overlap" creates a complex system of relationships and duties to manage. Wineries undertaking the wine tourism path must face pressures and challenges; more specifically, Lavandoski et al. (2016) talks about a different set of institutional pressures and organisational performances that characterise the two hemispheres of wine production and tourism. This further strengthens what we previously stated about the difficulties of managing activities that belong to "two clusters". Another exciting aspect should be underlined, and it deals with the issue of institutional pressure, a sense of commitment that an organisation experiences due to its belonging to a specific category of stakeholders.

It is fascinating that the degree of institutional pressure increases with the growth of professionalism across the companies: Organisations try to establish a set of rules that encompasses working methods and conditions to legitimate their professional autonomy.

The rising number of players in the wine tourism industry has improved the overall degree of institutional pressure.

2.8 Analysing Competitiveness

Wine tourism can contribute to a system's overall competitiveness or a single company. In general, the beneficial effects of wine tourism can have a micro and a macro impact. For what concerns the relationships between wine tourism and local economic development, we can say that it can be a pillar for the economic development of some areas: besides the role that wine tourism can have in revitalising some rural areas, it plays a pivotal role in attracting new investments, for example (Sanders, 2004). Wine tourism activities can create a positive circle for improving the awareness of a wine producing area.

Wineries can benefit directly from improved profits and incomes that derive from wine tourists and indirectly from the increased awareness of their brands if they can adequately manage the relationship with wine tourists (Koch et al., 2013). Wine tourism can be an education tool for visitors (Getz & Brown, 2006): through the planned activities, tourists gather information about the production process, the wine business, the cultural heritage that might influence the local wine production and local wine system, as well as the story of the winery and its products.

From another point of view, wine tourism can represent a tool for improving the distribution system developed by a winery since it can be considered a tool for direct distribution (Byrd et al., 2016; Getz & Brown, 2006) As outlined by Santos et al. (2021) wine tourism has been a wheel for solving some structural problems that affect the sector, such as a lack of infrastructures and resources for developing synergies between planned strategies that involve business, marketing and HR decisions (see also Hall & Mitchell, 2000).

A winery that decides to pursue this activity has to analyse its competitive environment's structure and evaluate the costs and benefits of entering the business.

In some cases, location can boost a company's competitiveness: background research has highlighted that the country of origin plays a pivotal role in choosing a destination in wine tourism (Orth et al., 2012). Countries with a strong brand can represent an asset for wineries, although a threat to the brand could compromise the competitive strategy pursued by the company. Companies should evaluate the degree of dependency of their wine tourism strategy on the territorial brand. They should also define how much of their competitiveness depends on their internal assets and how much from the external ones (territorial brand, country of origin).

We will outline some inputs that can be employed for assessing the industry scenario's competitiveness, which can affect the profitability of companies' choices.

Barriers to Entry and Potential Entrants

For companies that operate in the wine business, specifically for wineries, barriers to entry in the wine tourism industry are generally considered low (Carlsen, 2004). At the basis of this idea, there is the convincement that any synergy with the wine business helps wineries develop a wine tourism offer and plan experiences to offer, such as the winery and vineyard visit.

Barriers to entry are represented by the resources required by the activity and the industry. Researchers have explored the role of resources in shaping the competitive strategy of wineries that operate in wine tourism. Some studies (Guedes & Rebelo, 2019) have adopted a widely spread approach in the literature that focuses on the dichotomy of internal/external resources, underlining that wineries must deal not only with problems related to the emerging market demand (in terms of services to provide, for example) but also with issues that refer to the structure of the supply chain where wineries operate, that can be highly different from the traditional tourism supply chain. Internal resources are related to corporate's resources, and externals are related to the so-called contextual dimension. The degree of barriers to entry varies according to the characteristics of the winery and its location. When talking about location, we do not refer only to physical barriers that might represent an obstacle to reaching the winery or to establishing new buildings for the development of wine tourism but also to the presence of clustered organisations of wineries that can help other companies in making a wine tourism offer succeed.

According to Grimstad and Burgess (2014), small wineries might feel the pressures typical for small businesses: lack of time and knowledge for managing the new activity can reduce companies' chances to succeed. As well as for tourism Internet has altered the barriers to entry for players (Buhalis & Zoge, 2007), and it has improved the degree of transparency, lowering switching costs for consumers. The Internet has improved the chance to establish direct contact with visitors for producers and made the market more accessible in general. Thus, this has meant the development of new competencies and required companies to manage new activities. We are going back to what we have already outlined: smaller companies with limited capabilities and resources where owners have little time to manage visitors and establish direct contacts can be penalised in this scenario.

Profitability

Background research has highlighted that in the wine tourism business, following the main principles that regulate industry competition and business in general, companies can have the chance to establish the price, or in other words, they are price-makers.

The chance to set prices depends on various factors, such as the degree of horizontal competition that companies must face, the structure of the industry supply chain and the degree of vertical coordination (Nevo, 2001).

The growing number of wineries that develop similar offers depletes the chance to keep prices high unless companies provide services that motivate higher prices or can find ways to differentiate themselves.

Price is defined according to the provided service and visitors' perceived experience. Generally, the more experience wineries promise their visitors, the more they are entitled to ask for their services. It can also happen that the price settled for the visit reflects the perceived image of the winery and its products. Wineries that rely on a robust company brand can play higher prices since visitors would be attracted.

A clear example is the case of the Antinori Winery in the ChiantiShire in Tuscany. Antinori's headquarters is in an area rich in wineries, where wine tourism represents a relevant part of the business. Florence, one of Italy's leading tourist attractions, can easily reach the area. The winery belongs to the Antinori Family, and its product portfolio serves various market segments, including the very high-end market, where the company is recognised to play a leading role. From the website, the mix between history and tradition emerges: "A story that began in 1385. A historical link with the Chianti Classico region. A place where tradition and innovation coexist in perfect harmony. A winery that reveals the past, present and future of the Antinori family" (from the company website). The company opened the new winery in 2012, and the design is highly innovative: "In 2022, Antinori nel Chianti Classico winery was awarded first place in the World's Best Vineyards. This international ranking honours the most remarkable wine tourism destinations, voted by over 500 leading wine and tourism experts" (from the company website).

Some professionals and journalists well describe the reasons for visiting the winery:

> Antinori has huge historical significance in Italy's wine industry, playing a significant role in the 'Super Tuscan' revolution in the 70s. Back then, wine-makers were very tightly restricted in terms of the type and percentage of grape varietals that could be used in wine to get a certain accreditation, such as DOC or DOCG. Some wine-makers decided to ignore this classification and make 'Super Tuscans', wines from Tuscany made with blends and wine-making techniques that were not based on the conditions for accreditation. (Source: thetravellingsquid.com)

The company offers diversified tour packages and prices. Visitors can taste different types of wine, including those not produced near this cellar. The price for wine tour packages for Antinori is higher than the ones available in the local market, especially for basic wine tours. However, it is in line with the positioning company strategy that emerges from the description provided by the blogger and the mentioned.

Some exogenous factors—that influence price-making decisions—derive from the geographical context or a specific market structure that helps wineries set a certain price.

However, as the number of services increases, the cost of providing services grows, and mechanisms that erode profitability can occur.

Similarly, companies' ability to play as price-makers or, generally, to decide their pricing strategy gets lower when the number of intermediaries grows.

The number of intermediaries in wine tourism has evolved over the years because the business has boomed, and agencies and tour operators have started promoting wine tourism packages. ENIT states, "Food and wine tourism is the fourth best-selling travel and organised holiday product in Italy by 17.3% of the Tour Operators interviewed that sell products in Italy" (From ENIT Website).

Distribution of wine tourism offers can be direct or indirect: as it happens with wine, the wine tourism offer must reach a target customer and be distributed. Wineries must decide to reach the visitors by establishing a direct connection with prospective customers or hiring intermediaries to reach customers. The same considerations we have carried previously about the role of market structure and the capability of a company to differentiate itself and its production can work when discussing distribution. Furthermore, the global coverage of market requests for wine tourism has meant wineries to improve their presence on a global scale. Going global is a strategy: many companies have been attracted by the promising profitability of global high-end market niches, for example, or by the idea of reaching wine enthusiasts and tourists worldwide.

Another aspect that can reduce the profitability of wine tourism in the long run is that wineries should also consider the negative effects that wine tourism could have on the environmental resources that belong to the destination. In other words, a perspective that focuses on a long-term sustainable competitive advantage should be adopted in order not only to evaluate how long the competitive advantage will last but also how the chosen strategy is sustainable in terms of resource deployment (see, among the others Hassan, 2000).

As Hall and Mitchell (2000) have outlined, wineries must balance the pros and cons of wine tourism. On one side, they can help improve company brand awareness, reach new customers and improve profits and margins. Conversely, wineries must develop new competencies and capacities and attract capital to pursue goals and activities.

2.9 Discussion and Conclusions

Since the beginning of the chapter, a key message stands out: the balance between costs and benefits takes time to manage. Wineries, especially the small ones, often encounter a typical scenario. They are attracted by the opportunities the wine tourism business creates, seeing many potential visitors; then, they plan an offer focusing on their strengths, especially tradition and authenticity. Then, wineries realise that wine tourism has different rules from wine production. Although we have intentionally provided some general information about Tuscan wine tourism and Tuscan wineries, we can say that the case we have observed represents a starting point for a discussion that would focus on two main issues: the assets chosen for shaping the company's wine tourism offer and the dynamics that affect profitability in the evolving competitive scenario.

The title of the chapter is "Cornering Authenticity", and it recalls the idea of being trapped in a corner. Wineries can be trapped in a corner by the importance they give to an issue (authenticity in this case) or can trape it and give it limited space once they decide to emphasise other issues when planning a strategy.

The link with the tradition, the story of a territory and its production and the emphasis on cultural heritage, families and company history help promote an authentic wine tour experience. As we have underlined, authenticity can be an asset. However, it is a complex factor requiring deep investigation and reflection. The subjective perception of visitors underlines that the concept is not fixed but dynamic and that tourists' perceptions can change the image of what should be authentic. Thus, authenticity is an attribute of the experience, and, as well as other attributes, needs to be managed; furthermore, since we are talking about services, other attributes of the experience could be, for some visitors, more important than making an authentic experience and influence the overall perception of the experience.

It should be interesting to understand how the overall satisfaction of the entire wine tourism experience runs along the wine tour offer's entire process (pre-visit, visit and post-visit). Understanding how satisfaction changes according to the tourist's profile and the different factors contributing to creating a sense of satisfaction should also be helpful.

In the chapter, we talked about standards: some services are required, and that must be offered. In other words, it is essential to be professional. We can recall the concept of institutional pressure: given the growth of the business, the degree of professionalism has improved the institutional pressure, and it affects the company's profitability in the end.

As we have previously outlined, the decisional approach should highlight which resources (internal or external) support the competitiveness of the wine tourism offer. In the future, it would be interesting to compare Tuscan wineries in the wine tourism industry and non-wine tourism-oriented wineries to evaluate their competitiveness.

Some questions emerge: What is the degree of institutional pressure in wine tourism? How do we deal with institutional pressures? How can it become a stimulus for improving the quality of the offer, and under what circumstances can this represent a barrier?

Finally, we have considered the circular dynamism that could affect the new competitive scenario: we have referred to a circle, but it could become a dangerous loop. There could be a progressive growth of offered services to meet the needs of upgrading the offer. How long could this be sustainable? In this scenario, what is the role of assets, such as authenticity? The other dilemma for a winery is "Should the company follow the flow?". In other words, the real problem is deciding how much to be market-oriented and stay connected with traditional issues or…authenticity.

A solution could be focusing on the type of visitors, which creates a further need: companies would need data to profile actual and prospective visitors. How can they manage them? Do they have resources and competencies?

The number of questions that this chapter has raised is considerable. In our view, there is a need to investigate the implication of changes in the competitive scenario for wine tourism, especially in the light of small wineries. Further research should be carried out in this direction.

References

Barbierato, E., Bernetti, I., & Capecchi, I. (2021). Analyzing TripAdvisor reviews of wine tours: An approach based on text mining and sentiment analysis. *International Journal of Wine Business Research, 34*(2), 212–236.

Bruwer, J. (2003). South African wine routes: Some perspectives on the wine tourism industry's structural dimensions and wine tourism product. *Tourism Management, 24*(4), 423–435.

Bruwer, J., & Lesschaeve, I. (2012). Wine tourists' destination region brand image perception and antecedents: Conceptualization of a winescape framework. *Journal of Travel & Tourism Marketing, 29*(7), 611–628.

Bruwer, J., Prayag, G., & Disegna, M. (2018). Why wine tourists visit cellar doors: Segmenting motivation and destination image. *International Journal of Tourism Research, 20*(3), 355–366.

Buhalis, D., & Zoge, M. (2007). The strategic impact of the Internet on the tourism industry. In M. Sigala, L. Mich, & J. Murphy (Eds.), *Information and communication technologies in tourism 2007* (pp. 481–492). Springer.

Byrd, E. T., Canziani, B., Hsieh, Y. C. J., Debbage, K., & Sonmez, S. (2016). Wine tourism: Motivating visitors through core and supplementary services. *Tourism Management, 52*, 19–29.

Carlsen, P. J. (2004). A review of global wine tourism research. *Journal of Wine Research, 15*(1), 5–13.

Charters, S. (2006). Aesthetic products and aesthetic consumption: A review. *Consumption, Markets and Culture, 9*(3), 235–255.

de Freitas Coelho, M., de Sevilha Gosling, M., & de Almeida, A. S. A. (2018). Tourism experiences: Core processes of memorable trips. *Journal of Hospitality and Tourism Management, 37*, 11–22.

Dressler, M., & Paunovíc, I. (2021). Business model innovation: Strategic expansion of German small and medium wineries into hospitality and tourism. *Administrative Sciences, 11*(4), 146.

Getz, D., & Brown, G. (2006). Critical success factors for wine tourism regions: A demand analysis. *Tourism Management, 27*(1), 146–158.

Grimstad, S., & Burgess, J. (2014). Environmental sustainability and competitive advantage in a wine tourism micro-cluster. *Management Research Review, 37*(6), 553–573.

Guedes, A., & Rebelo, J. (2019). Merging wine and tourism-related services: Evidence from the Douro (Portugal) Wine Region. *Journal of Wine Research, 30*(4), 259–274.

Hall, C. M., & Mitchell, R. (2000). Wine tourism in the Mediterranean: A tool for restructuring and development. *Thunderbird International Business Review, 42*(4), 445–465.

Haller, C., Thach, L., & Olsen, J. (2020). Understanding eWinetourism practices of European and North America wineries. *Journal of Gastronomy and Tourism, 4*(3), 141–156. ISO 690.

Hassan, S. S. (2000). Determinants of market competitiveness in an environmentally sustainable tourism industry. *Journal of Travel Research, 38*(3), 239–245.

Jamal, T., & Hill, S. (2004). Developing a framework for indicators of authenticity: The place and space of cultural and heritage tourism. In *The political nature of cultural heritage and tourism* (pp. 211–230). Routledge.

Kim, H., & Bonn, M. A. (2016). Authenticity: Do tourist perceptions of winery experiences affect behavioral intentions? *International Journal of Contemporary Hospitality Management, 28*(4), 839–859.

Koch, J., Martin, A., & Nash, R. (2013). Overview of perceptions of German wine tourism from the winery perspective. *International Journal of Wine Business Research, 25*(1), 50–74.

Kovačić, S., Jovanović, T., Vujičić, M. D., Morrison, A. M., & Kennell, J. (2022). What shapes activity preferences? The role of tourist personality, destination personality and destination image: Evidence from Serbia. *Sustainability, 14*(3), 1803.

Lavandoski, J., Pinto, P., Silva, J. A., & Vargas-Sánchez, A. (2016). Causes and effects of wine tourism development in wineries: The perspective of institutional theory. *International Journal of Wine Business Research, 28*(3), 266–284.

Mehmetoglu, M. (2007). Typologising nature-based tourists by activity—Theoretical and practical implications. *Tourism Management, 28*(3), 651–660.

Morpeth, N. (2009). Diversifying wine tourism products: An evaluation of linkages between wine and cycle tourism. In C. M. Hall, L. Sharples, B. Cambourne, N. Macionis (Eds.), *Wine tourism around the world* (pp. 272–282). Routledge.

Nevo, A. (2001). Measuring market power in the ready-to-eat cereal industry. *Econometrica, 69*(2), 307–342.

Nowak, L. I., & Newton, S. K. (2006). Using the tasting room experience to create loyal customers. *International Journal of Wine Marketing, 18*(3), 157–165.

Orth, U. R., Stöckl, A., Veale, R., Brouard, J., Cavicchi, A., Faraoni, M., Larreina, M., Lecat, B., Olsen, J., Rodriguez-Santos, C., Santini, C., & Wilson, D. (2012). Using attribution theory to explain tourists' attachments to place-based brands. *Journal of Business Research, 65*(9), 1321–1327.

Overend, D. (2012). Performing sites: Illusion and authenticity in the spatial stories of the guided tour. *Scandinavian Journal of Hospitality and Tourism, 12*(1), 44–54.

Plasait, B. (2007). *Le tourisme d'affaires: un atout majeur pour l'économie*. Les éditions des journaux officiels.

Priilaid, D., Ballantyne, R., & Packer, J. (2020). A "blue ocean" strategy for developing visitor wine experiences: Unlocking value in the Cape region tourism market. *Journal of Hospitality and Tourism Management, 43*, 91–99.

Sanders, D. (2004). Wine tourism and regional development: A case study of the Margaret River region, Western Australia. In *International wine tourism conference* (pp. 1–13). Vineyard Publishing.

Santini, C. (2019). The business of wine tourism: Evolution and challenges. In M. Sigala & R. Robinson (Eds.), *Management and marketing of wine tourism business: Theory, practice, and cases* (pp. 261–276). Palgrave Macmillan.

Santini, C., & Faraoni, M. (2010). Wine clubs and direct sales: Learning a lesson from the USA. *Managing the Wine Business: Research Issues and Cases*, 197–208.

Santos, V. R., Ramos, P., Almeida, N., & Santos-Pavón, E. (2019). Wine and wine tourism experience: A theoretical and conceptual review. *Worldwide Hospitality and Tourism Themes, 11*(6), 718–730.

Santos, V., Ramos, P., Sousa, B., & Valeri, M. (2021). Towards a framework for the global wine tourism system. *Journal of Organizational Change*

Management, *35*(2), 348–360. https://doi.org/10.1108/JOCM-11-2020-0362

Su, L., Tang, B., & Nawijn, J. (2021). How tourism activity shapes travel experience sharing: Tourist well-being and social context. *Annals of Tourism Research, 91*, 103316.

Websites

https://www.enit.it/storage/202302/20230201171304_221115%20report%20t.o.%20focus%20enogastronomia%20isnart-enit.pptx_it-it_en-gb.pdf

https://www.nomisma.it

https://www.regione.toscana.it/documents/10180/15105612/I+numeri+del+vino+in+Toscana+Studio+ISMEA_03.2022.pdf/b62358b3-1c8d-342e-db82-1934faf692d1?t=1648540698157#:~:text=La%20più%20recente%2C%20fotografia%20della,7%25%20rispetto%20alla%20campagna%20precedente

https://www.tuscanywineries.it/en/home-en/

https://www.i-tuscany.com

https://www.diacceroni.com

3

Knowledge Management, Innovation and Business Growth in Wine Tourism: The Case of *Finca La Estacada*

Jorge Linuesa Langreo, Pablo Ruiz-Palomino, Rafael Morales Sánchez, and María Ángeles García-Haro

3.1 Introduction

The phenomenon of wine tourism has been growing in recent years, driven by the interest of tourists in discovering the secrets of the world of wine (Hall et al., 2009). This has provided wine firms with a tool to improve their viability and sustainability by establishing synergies

J. L. Langreo (✉) · P. Ruiz-Palomino · M. Á. García-Haro
Department of Business Administration, University of Castilla-La Mancha, Cuenca, Spain
e-mail: jorge.linuesa@uclm.es

P. Ruiz-Palomino
e-mail: pablo.ruiz@uclm.es

M. Á. García-Haro
e-mail: Angeles.Garcia@uclm.es

R. M. Sánchez
Department of Management and Marketing, Universidad Pablo de Olavide, Seville, Spain
e-mail: rmorsan@upo.es

J. Martínez-Falcó et al. (eds.), *Strategic Management in the Wine Tourism Industry*,
https://doi.org/10.1007/978-3-031-54837-6_3

between tourism and wine production (Trigo & Silva, 2022). In this way, wine tourism has emerged as a robust strategy to diversify wine production, adding value to the tourism offer with the incorporation of exclusive services and experiences (Croce & Perri, 2017). Accordingly, tourist destinations and, consequently, wineries, have been obliged to evolve and adapt to the demands of tourists, which has led to the need for reinvention and transformation of the wineries themselves and their management, incorporating spaces and activities that encourage, attract and promote wine tourism demand (Hall et al., 2009).

In general terms, a firm's ability to innovate is a critical aspect of its growth strategy (Grant, 2021), which depends on the correct management of organizational knowledge (Gómez & Manzanares, 2006). In other words, how a firm efficiently acquires, creates, manages and reconfigures knowledge (both inside and outside the organization) will determine its capacity to adapt to changes in the environment (Hussi, 2004). In this way, knowledge is an essential competitive resource in generating a continuous flow of innovations that allows a firm to adapt to the different needs of its customers (Leidner et al., 2006).

This case study analyzes how *Finca La Estacada* has developed a diversification strategy based on an optimal management of organizational knowledge that has allowed it to generate the necessary innovations to adapt to market changes. For this analysis, we first present the main aspects of wine tourism. Next, the relationship between organizational knowledge, innovation and growth is analyzed. Subsequently, the case of *Finca La Estacada* is studied, examining, on the one hand, its formation and growth and, on the other, its knowledge management and innovation strategy. Finally, some practical implications are detailed and conclusions are drawn on how knowledge management can be a key factor in business growth strategy.

3.2 Wine Tourism

Wine tourism is defined as "trips and stays aimed at getting to know the landscapes, the work and the spaces where wine is made, and the activities that increase knowledge and acquisition and can generate development in the different wine-growing areas" (Pastor & Vicente, 2006, p. 64). In this way, wine represents the essence of wine tourism, and is what allows the tourist to enjoy a unique experience linked to a specific space and time, since each vintage is different, with distinctive and incomparable wines (Favero & Rotta, 2007). Therefore, tourism linked to grape culture has become an exclusive tourism product that combines leisure, culture and the enjoyment of wine around a unique enological heritage (Hall et al., 2009).

Enological heritage is defined as "the set of spaces, objects, actions and oral references related to wine activity" (Pastor & Vicente, 2006, p. 106). Specifically, Enological heritage is composed of all the resources and capacities necessary to carry out the production and elaboration of wine. Due to its importance, it is possible to elaborate an ecological heritage inventory incorporating all the elements that generate value for the winery, which allow for the development of the winemaking activity, and the design of tourism promotion. Following Pastor and Vicente (2006), we can identify five essential elements: (a) the *space*, referring to the land, the vineyard, the landscape and the existing natural resources; (b) the *field work*, including all the elements necessary for the planting of the vine, its cultivation and harvesting; (c) the *winery work*, being the main space where the winemaking and wine tourism-related activities are carried out; (d) the existing *buildings and facilities* in the wine tourism complex; (e) the *intangible heritage*, which includes oral tradition, related legends, social, cultural and religious aspects, iconography and the culture associated with wine, among other aspects.

All these differential characteristics give the winery a unique character as it is where the entire value chain process takes place, from the cultivation, production and elaboration of the wine to the provision of an exclusive service to the tourist visiting the winery or the region (Hall et al., 2009). For all these reasons, we speak of heritage resources related to a winery or a territory, as tourists link them to the indigenous cultivation of a region or a winery, which offers fruits, products and services that reflect the culture, customs and tradition of that place (Mitchell & Hall, 2006). Ultimately, wine tourism also plays an important role within wine-producing territories as it allows for the preservation and dissemination of culture and traditions around wine activity (Croce & Perry, 2017).

The development of tourism related to wine heritage has been evolving and growing over time due, among other reasons, to the segmentation of tourism and the consequent specialization of tourism products and experiences (Koch et al., 2013). Today's tourists demand a wide range of exclusive and personalized products from wineries. It is therefore necessary for wineries to define and delimit the profile of the tourist visiting the winery and the motivations that lead them to do so (Charters & Ali-Knight, 2002). In other words, it is a matter of framing the different tourist profiles in order to analyze their needs and thus adapt the offer to their demands (Li et al., 2017). This is why the socio-demographic characteristics of visitors, the number of days spent at the destination, their knowledge of wine or the way in which they learned about the destination are key aspects for analyzing and forecasting demand and, thus, also for adjusting the supply.

In short, although the main objective of wineries is to produce and market wine, they see wine tourism as a highly valuable complementary resource where they can diversify their activity (Hall et al., 2009). In order to develop a synergistic effect between wine production and wine tourism, it is necessary to generate, among tourists, the desire or motivation to travel to the place where this exclusive product is produced and which they have previously tasted (Marzo-Navarro & Pedraja-Iglesias, 2012). In other words, there must be a clear association between the image of the products marketed by the winery and the tourist destination (Rico, 2011). To achieve this, wineries need to develop an adaptation process that, in general terms, involves (Rico, 2011): (a) setting up the facilities, (b) identifying the needs of wine visitors, (c) implementing loyalty programs, (d) having staff training plans and (e) improving general professionalization in management and leadership. The idea is to create an "Open Winery" model (Badulescu & Badulescu, 2016) that offers tourists a unique experience through guided tours of the wineries and vineyards, wine tasting and an extensive gastronomic and cultural offer. In short, wineries can channel all their knowledge of the sector toward the development of a model of continuous innovation around wine tourism.

3.3 Organizational Knowledge, Innovation and Growth

In the current scenario, characterized by a high degree of uncertainty, knowledge has become a basic strategic resource for a firm, allowing it to obtain competitive advantages (Grant, 2021). The knowledge that firms accumulate not only allows them to be much more efficient, but also enables them to generate new opportunities for innovation and growth (Miles et al., 1997). In fact, the generation of new products and services is based on the understanding of the synergies and interactions between a firm's knowledge and the different market opportunities presented to them (Helfat & Raubitschek, 2000). Therefore, appropriate knowledge management allows a firm to direct its resources and capabilities toward innovation, growth and achieving organizational objectives (Martínez-Pérez et al., 2016).

Studies such as those developed by Nonaka (1991) or Grant (1996) consider organizations as knowledge-based systems, with this being a key resource for swiftly responding to changes in the environment and taking advantage of business opportunities, while also influencing innovation and business growth (Grant, 1996; Teece et al., 1997). These aspects are aligned with the theory of resources and capabilities (Barney et al., 2011), as studies in this field consider that the creation, accumulation and application of knowledge may be the origin of the differences observed in innovation and business growth in firms in the same sector (Martínez-Pérez et al., 2016). Therefore, the organizational knowledge management strategy followed by a firm will be an essential element when it comes to generating different innovations and competitive advantages.

Knowledge management has traditionally been divided into knowledge exploration and exploitation (Martínez Pérez et al., 2016). On the one hand, knowledge exploitation refers to the development of resources and capabilities that allows a firm to take advantage of the existing knowledge within the organization (Gobbo & Olsson, 2010). Such exploitation thus reflects the management of existing knowledge in its operations that are aimed at improving the efficiency of the organization's resources and capabilities and succeeding in creating new combinations (Levinthal & March, 1993). On the other hand, knowledge exploration focuses on the development of resources and capabilities for the creation or acquisition of new knowledge and the improvement of current organizational knowledge. It is associated with experimentation and the search for new knowledge that is external to the organization (Gobbo & Olsson, 2010). All this leads a firm to become more innovative and competitive in the face of market changes (March, 1991).

The ability of organizations to exploit their knowledge has a decisive influence on the generation of innovations (Rabeh et al., 2013). This is because, in many cases, innovations are based on new ways of combining elements and pre-existing knowledge in the organization (Nahapiet & Ghoshal, 1998). Thus, a firm's ability to generate innovation depends on the effort it makes in reinforcing the knowledge it possesses (Subramaniam & Youndt, 2005). Thus, if organizations pursue an exploitative knowledge strategy, they must invest in practices that facilitate human resource management, improve customer satisfaction, promote a corporate culture and leadership that supports innovation and continuous improvement.

In addition, in order to generate innovations, firms must develop a parallel strategy of knowledge exploration (Gobbo & Olsson, 2010). That is, the generation of new knowledge that goes beyond existing knowledge and is associated with experimentation, flexibility and divergent thinking (Levinthal & March, 1993). In order to find new solutions, organizations encourage the acquisition of knowledge from outside the organization, which will enable new combinations of organizational knowledge and new possibilities for innovation (Ahuja & Lampert, 2001). Thus, an organization's ability to generate innovation will greatly depend on its access to knowledge that did not previously exist; that is, the development of an exploratory knowledge strategy (Subramaniam & Youndt, 2005). In this way, the development of cooperation agreements or collaborative projects with other firms and institutions can be an essential element for the acquisition of external knowledge and the generation of new innovations (Martínez-Costa et al., 2019).

In summary, knowledge management (exploitation and exploration) contributes to the generation of higher levels of innovation and growth in a firm since: (a) the firm is able to generate innovations and develop internally from exploited knowledge, (b) the firm acquires external knowledge that it can integrate within the organization, (c) the organization incorporates knowledge through cooperation agreements or collaborative projects with other organizations or institutions. The factors that have affected the growth process of *Finca La Estacada* are analyzed below.

3.4 Wine Tourism at *Finca La Estacada*

The Formation and Growth of *Finca La Estacada*

Finca La Estacada is located in Tarancón, actually a municipality in the Castilla-La Mancha province of Spain. Tarancón is located about 45 minutes away from Madrid and it has approximately around 16,200 inhabitants. It is well connected to various cities and municipalities through highways and railways. Tarancón stands out for having some of the best wineries in the autonomous community and being able to enjoy the Uclés Wine Route and the enotourism experience.

The origin of the wine tourism complex can be traced back to the Cantarero family's acquisition, in the mid-twentieth century, of an old estate owned by the descendants of Queen María Cristina, widow of Fernando VII. The establishment takes its name from the site where it is located, called *La Estacada*, of Roman origin. It was formerly the site of the road that connected Segóbriga to Completum (now Alcalá de Henares).

The winery is located on a 5-hectare plot called "*La Estacada*". Outside, it has twelve 50,000 L fermentation tanks, next to which all the harvesting machinery is located. The estate adapts to the surrounding vineyards, integrating itself into them, while also adapting its facilities to the landscape. Additionally, it is one of the most innovative and technologically advanced wineries in Spain.

Finca La Estacada currently owns 278 hectares of vineyards located in the vicinity of the winery's facilities, where one can find varieties such as Tempranillo, Cabernet, Merlot, Syrah, Mazuelo, Prieto, Picudo, Petit Verdot, Malbec, Graciano, Viognier and Sauvignon Blanc. In addition, its facilities include an aging room with 3,000 barrels of American and French oak, which are located in the central building of the winery, in an underground warehouse of 3,000 m^2. Its being underground means that the temperature and humidity in the room remain constant throughout the year.

Currently, *Finca La Estacada* sells more than twenty wines in the national and international market, being present in the five continents and reaching countries as diverse as Switzerland, Poland, Serbia, China, Taiwan, Japan, Turkey, the United States, Colombia, Costa Rica, Australia and New Zealand, among others.

The company Finca La Estacada has 30 employees, and it is ranked 273rd in the Wine Production sector in Spain (see Figs. 3.1 and 3.2).

As can be seen in the Fig. 3.3, the sales for the last registered year, 2022, amount to 3,906,214.28 euros, which is nearly 13% more than the previous year, with a total asset of 20,707,240.26 euros. If measured in terms of the company's sales contribution to the GDP of its province, the company contributes 0.09% to the local GDP of Cuenca. It's worth noting that the company has won the TripAdvisor 2023 Travelers' Choice

Fig. 3.1 Summary of positions at *Finca La Estacada* (*Source* Adapted from eInforma. https://ranking-empresas.eleconomista.es/)

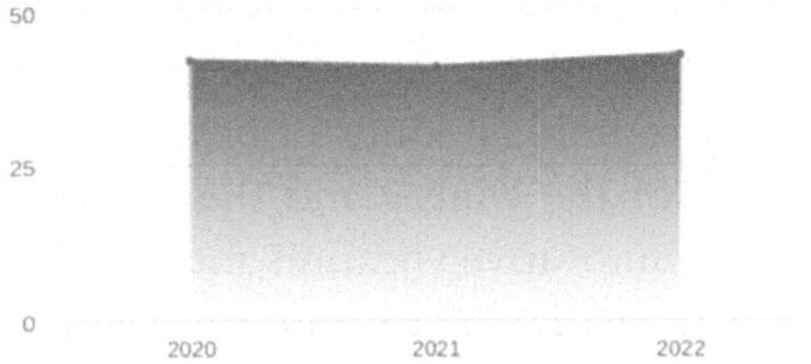

Year	Total	Permanent employees	Temporary employees
2020	42	85.71	14.28
2021	41	78.04	21.95
2022	43	100.00	

Fig. 3.2 Employee evolution of *Finca La Estacada* (*Source* Adapted from https://www.einforma.com/)

Award, becoming one of the favorite places for travelers, as evidenced by the large number of excellent reviews on the platform. Specifically, the hotel ranks among the top 10% in the world.

Although wine production is at the heart of *Finca La Estacada*, the wine tourism complex offers a wide range of services to tourists. Firstly, it has a restaurant that is highly spacious and luminous thanks to its vaults and large windows from which extensive vineyards surrounding

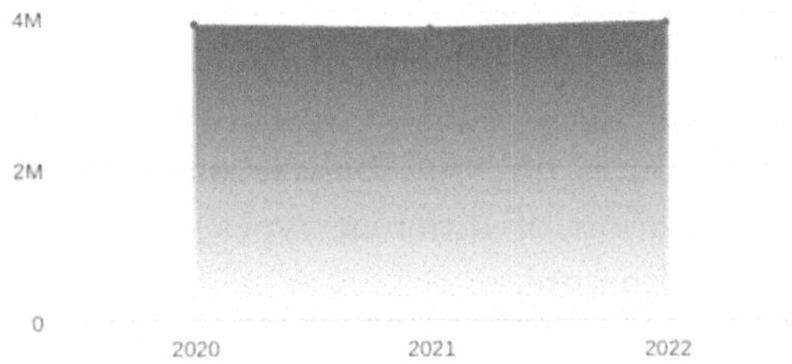

Year	Value (€)
2020	3,857,248.44
2021	3,815,645.74
2022	3,906,214.28

Fig. 3.3 Sales Evolution of *Finca La Estacada* (*Source* Adapted from https://www.einforma.com/)

the winery can be enjoyed, and which has direct access to the barrel room. In addition, it has a large terrace located on the porch of the winery, which allows it to hold all kinds of presentations, events or tastings of both the wines of the winery and the dishes of the restaurant. Finally, as a result of its innovative vocation and its high customer orientation, the *enobar* was created, in 2016, as a new space integrated into the restaurant, where customers can enjoy the wide range of wines produced by the winery.

The facilities were extended in 2009 with the creation of a hotel consisting of 25 fully equipped rooms adapted to the different needs of customers. In addition, the same building houses a wine therapy spa so that tourists can enjoy the properties and benefits of the wines, not only through sight, smell and taste, but also through the skin.

The wine tourism complex is completed with: (a) two tasting rooms, (b) a shop, where not only the winery's wines are sold, but also the typical products of the region's gastronomy and (c) an informative classroom on wine culture, where visitors can also observe in situ, via an audiovisual presentation, the entire winemaking process from the emergence of the grape to the marketing of the bottle ready for sale.

The firm commitment of *Finca La Estacada* to wine tourism has led it to design complementary activities that combine innovation and tradition. In this way, activities related to gastronomy, cultural heritage or events are carried out, seeking to promote the value of the wine tourism complex as an alternative tourist resource. Examples of these include conventions, guided visits to the winery, tastings and samplings where tourists can come into direct contact with the product. Others range from the design of routes on foot, bicycle or horseback through the estate, to the organization of different types of events of a business, cultural or recreational nature.

The Strategy of Knowledge Management and Innovation at *Finca La Estacada*

The growth strategy of *Finca La Estacada* is based on the knowledge generated by the organization itself. Drawing on its knowledge of wine growing, it has managed to diversify its activity toward the development of new products and services linked to wine tourism with the aim of providing a complete experience for tourists (catering, accommodation and complementary services). In this way, the firm has succeeded in diversifying its business activity, dedicating itself not only to enological activity, but also adapting to the offer of services to meet tourist demand. Through its knowledge strategy, *Finca La Estacada* has integrated tourism activities into the production and marketing of its wines. Moreover, its economic and financial situation is sound and it has experienced a period of international growth, with exports of its products accounting for more than 60% of its turnover. All these aspects have allowed it to consolidate the *Finca La Estacada* brand and generate a synergistic effect between production and wine tourism. It is currently facing new growth challenges, namely, to increase its position in the wine tourism industry and to strengthen its activity in international markets.

One of the main factors enabling the growth of *Finca La Estacada* and its involvement in the different knowledge management activities is its corporate culture. The culture of *Finca La Estacada* has a strong orientation toward knowledge creation and innovation, which is supported by elements such as the promotion of an open and honest communication between all the members of the firm, organizational flexibility and the commitment of all the employees to innovation.

Together with the organizational culture, its human resources policy is another of the pillars on which its growth strategy has been based. *Finca La Estacada* boasts highly qualified employees, who are organized into different working groups that collaborate on a daily basis to promote the creation and distribution of knowledge and the development of innovations. At the same time, *Finca La Estacada* has implemented a continuous training plan for its employees, which allows them to update their knowledge in their daily practice and put it at the service of the organization. In addition, the managers engage in participative leadership with their employees, thus fostering high levels of motivation and commitment to the firm's future.

Finally, a third element that has allowed for the reinforcement of the firm's knowledge strategy and business growth is the cooperation agreements or collaboration projects that *Finca La Estacada* has established with other firms and institutions, which have allowed it to acquire the external knowledge necessary to be able to adapt to the different changes in the market.

3.5 Conclusions

Recent years have seen a change in the mentality of tourists and in their demands as they seek new experiences that lead them away from mass tourism destinations. Faced with this change in tourists' motivations, firms are developing new types of tourism that go beyond the traditional sun and beach type. These seek to respond to the demands of tourists eager to explore and discover the local customs of the destinations they visit, as well as their gastronomy and, consequently, their wines (Pastor & Vicente, 2006). Thus, wine tourism is not only a way for wineries to grow and diversify their products, but it has also become an essential element in the tourism and socio-economic development of many rural destinations. The commitment to this type of tourism translates into an increase in the number of visits to the wineries and the existing tourist offer in the area. This is why wineries should consider wine tourism as a complementary activity to the marketing and production of their wines, as well as being a multiplying factor in the socio-economic growth of the region.

This is the case of *Finca La Estacada*, which has managed to cement its growth through a clear commitment to wine tourism. The success of this growth can be explained by a clear commitment to a knowledge strategy that has allowed it to generate a continuous flow of innovations aimed at providing a high-quality service to customers. *Finca La Estacada* ensures that tourists, from the moment they enter the wine tourism complex, are provided with all the necessary services to enjoy a unique wine-related experience: accommodation, restaurants, spa, guided tours, wine therapy sessions, leisure and cultural activities, etc. The fundamental aspect that makes this whole process possible is the knowledge management approach as a global strategy that involves all the members of the firm through the promotion of an organizational culture oriented toward the generation of knowledge and innovation. Also key is the development of human resources policies that allow all its members to have the flexibility and communication necessary to provide this exclusive service. Furthermore, the ability of *Finca La Estacada* to establish cooperation agreements with other firms and institutions allows them to obtain the complementary external knowledge required to adapt to market changes.

In summary, although each firm must implement its strategy based on a personalized internal and external analysis, the present case shows how to confront the rapid changes occurring in the wine industry, especially in firms that have opted for wine tourism. It is necessary to establish a clear growth strategy based on the appropriate management of organizational knowledge and the creation of innovations aimed at providing a comprehensive and high-quality service to tourists.

References

Ahuja, G., & Morris Lampert, C. (2001). Entrepreneurship in the large corporation: A longitudinal study of how established firms create breakthrough inventions. *Strategic Management Journal, 22*(6–7), 521–543.

Badulescu, A., & Badulescu, D. (2016). Entrepreneurship and local resources. In *Tourism enterprises and the sustainability agenda across Europe* (pp. 169–186). Routledge.

Barney, J. B., Ketchen, D. J., Jr., & Wright, M. (2011). The future of resource-based theory: Revitalization or decline? *Journal of Management, 37*(5), 1299–1315.

Charters, S., & Ali-Knight, J. (2002). Who is the wine tourist? *Tourism Management, 23*(3), 311–319.

Company Profile Report "Finca La Estacada". https://www.einforma.com/

Croce, E., & Perri, G. (2017). *Food and wine tourism.* Cabi.

Fávero, I. R., & Rotta Antunes, J. (2007). Ecoturismo en la Regi6n de la uva y el vino: Brasil. *Estudios y Perspectivas En Turismo, 16*(2), 133–149.

Gobbo, J. A., Jr., & Olsson, A. (2010). The transformation between exploration and exploitation applied to inventors of packaging innovations. *Technovation, 30*(5–6), 322–331.

Gómez, F. G., & Manzanares, M. J. D. (2006). Conocimiento organizativo, innovación y crecimiento empresarial: al caso del Grupo Tecnobit. *Universia Business Review, 12*, 50–65.

Grant, R. M. (1996). Toward a knowledge-based theory of the firm. *Strategic Management Journal, 17*(S2), 109–122.

Grant, R. M. (2021). *Contemporary strategy analysis*. Wiley.

Hall, C. M., Johnson, G., Cambourne, B., Macionis, N., Mitchell, R., & Sharples, L. (2009). Wine tourism: An introduction. In *Wine tourism around the world* (pp. 1–23). Routledge.

Helfat, C. E., & Raubitschek, R. S. (2000). Product sequencing: Co-evolution of knowledge, capabilities and products. *Strategic Management Journal, 21*(10–11), 961–979.

Hussi, T. (2004). Reconfiguring knowledge management-combining intellectual capital, intangible assets and knowledge creation. *Journal of Knowledge Management, 8*(2), 36–52.

Koch, J., Martin, A., & Nash, R. (2013). Overview of perceptions of German wine tourism from the winery perspective. *International Journal of Wine Business Research, 25*(1), 50–74.

Leidner, D., Alavi, M., & Kayworth, T. (2006). The role of culture in knowledge management: A case study of two global firms. *International Journal of e-Collaboration, 2*(1), 17–40.

Levinthal, D., & March, J. (1993). The myopia of learning. *Strategic Management Journal, 14*(S2), 95–112.

Li, Y., Hu, C., Huang, C., & Duan, L. (2017). The concept of smart tourism in the context of tourism information services. *Tourism Management, 58*, 293–300.

March, J. G. (1991). Exploration and exploitation in organizational learning. *Organization Science, 2*(1), 71–87.

Martínez-Costa, M., Jiménez-Jiménez, D., & Dine Rabeh, H. A. (2019). The effect of organisational learning on interorganisational collaborations in innovation: An empirical study in SMEs. *Knowledge Management Research & Practice, 17*(2), 137–150.

Martínez-Pérez, Á., García-Villaverde, P. M., & Elche, D. (2016). The mediating effect of ambidextrous knowledge strategy between social capital and innovation of cultural tourism clusters firm. *International Journal of Contemporary Hospitality Management, 28*, 1484–1507.

Marzo-Navarro, M., & Pedraja-Iglesias, M. (2012). Critical factors of wine tourism: Incentives and barriers from the potential tourist's perspective.

International Journal of Contemporary Hospitality Management, 24(2), 312–334.

Miles, R. E., Snow, C. S., Mathews, J. A., Miles, G., & Coleman, H. J., Jr. (1997). Organizing in the knowledge age: Anticipating the cellular form. *Academy of Management Perspectives, 11*(4), 7–20.

Mitchell, R., & Hall, C. M. (2006). Wine tourism research: The state of play. *Tourism Review International, 9*(4), 307–332.

Nahapiet, J., & Ghoshal, S. (1998). Social capital, intellectual capital, and the organizational advantage. *Academy of Management Review, 23*(2), 242–266.

Nonaka, I. (1991). The knowledge-creating company. *Harvard Business Review, 69*, 96–104.

Pastor, E., & Vicente, L. (2006). *El turismo del vino. Documentos de estudios de ocio*. Publicaciones de la Universidad de Deusto.

Rabeh, A. D. H., Jimenéz-Jimenéz, D., & Martínez-Costa, M. (2013). Managing knowledge for a successful competence exploration. *Journal of Knowledge Management, 17*(2), 195–207.

Rico, M. G. (2011). *El turismo enológico desde la perspectiva de la oferta*. Ramón Areces.

Subramaniam, M., & Youndt, M. A. (2005). The influence of intellectual capital on the types of innovative capabilities. *Academy of Management Journal, 48*(3), 450–463.

Teece, D. J., Pisano, G., & Shuen, A. (1997). Dynamic capabilities and strategic management. *Strategic Management Journal, 18*(7), 509–533.

Trigo, A., & Silva, P. (2022). Sustainable development directions for wine tourism in Douro Wine Region, Portugal. *Sustainability, 14*(7), 3949.

4

Coopetition Strategies and Wine Tourism Offerings

James M. Crick and David Crick

4.1 Introduction

Within the cross-disciplinary literature underpinning this investigation, coopetition can be viewed as a mindset among decision-makers (Brandenburger & Nalebuff, 1996; Crick & Crick, 2024). However, earlier studies typically feature coopetition behavior in terms of value-creation activities arising from owner-managers cooperating with their competitors (Bengtsson & Kock, 1999; Bengtsson et al., 2016; Bouncken et al., 2020; Crick et al., 2021, 2023b; Czakon et al., 2020a, 2020b). It is "a paradoxical relationship between two or more actors, regardless of

J. M. Crick
School of Business, University of Leicester, Leicester, UK
e-mail: james.crick@leicester.ac.uk

D. Crick (✉)
Telfer School of Management, University of Ottawa, Ottawa, ON, Canada
e-mail: dcrick@uottawa.ca

J. Martínez-Falcó et al. (eds.), *Strategic Management in the Wine Tourism Industry*,
https://doi.org/10.1007/978-3-031-54837-6_4

whether they are involved in horizontal or vertical relationships, simultaneously in cooperative and competitive interactions" (Bengtsson & Kock, 2014: 180).

Coopetition strategies have been studied in the capacity of the global wine sector, in terms of vineyards and wineries working with their competitors (formally and informally) to achieve objectives that would be less likely if they were to operate with their own resources and capabilities (see Crick, 2019; Crick & Crick, 2023a; Crick et al., 2024; Dana et al., 2013; Granata et al., 2018). In other words, evidence has shown that wine producers (e.g., those that are focused on wine tourism offerings) are often well-served by collaborating with their competitors, since these networks can lead (if implemented correctly) to them yielding higher levels of organizational performance (Crick & Crick, 2021a, 2021b; Telfer, 2000).

That said, there are some research gaps pertaining to coopetition strategies in wine regions (including how coopetition links with wine tourism offerings). Specifically, the nature of how service-oriented organizations can manage these business-to-business activities is decidedly under-researched, as it is unclear how entrepreneurs can develop their wine tourism offerings by working with their competitors (e.g., creating value for their customers with augmented products and services). Accordingly, under the wider themes of resource-based theory (following Barney, 2018; Lavie, 2006), the objective of this chapter is to examine how coopetition strategies can be used to facilitate wine tourism strategies. In doing so, the following three contributions are made:

1. Stronger evidence emerges on the intricacies of how decision-makers (within vineyards and wineries—focused on delivering wine tourism offerings) can collaborate with their rivals to boost their market-level performance (extending Crick et al., 2022b; Telfer, 2000).
2. Improved insights are offered on the potential dark-sides of coopetition, namely, the organizational tensions that could be yielded from wine producers cooperating with their rivals in the capacity of wine tourism offerings (developing Crick & Crick, 2021a; Crick et al., 2022a).

3. New light is shed on the wider aspects of resource-based theory via highlighting how coopetition can be used by owner-managers (here, in vineyards and wineries) to survive and prosper within their markets (building upon Barney, 2018; Lavie, 2006).

To fulfill these novel contributions, the remaining sections of this chapter are presented as follows. First, the extant literature is reviewed to develop a conceptual model. Second, some practitioner implications are outlined. Third, the chapter is concluded, coupled with a series of future research directions.

4.2 Literature Review

Resource-Based Theory

Resource-based theory highlights that entrepreneurs can boost firm performance through effectively managing their assets, namely, resources and capabilities (Crick & Crick, 2021a). In other words, well-managed companies of different sizes under the value, rarity, inimitability, and non-substitutional (VRIN) criteria, can yield higher degrees of firm performance (Barney, 1991). This is achieved by effectively employing tangible assets (resources, like cash and hardware) and intangible assets (capabilities, such as knowledge and education) to out-perform their competitors. Nevertheless, recognition exists that performance, as an outcome of decision-makers' strategies, can be measured in varying objective and subjective ways (Bouncken & Fredrich, 2012; Crick & Crick, 2020; Estrada & Dong, 2020; Kraus et al., 2019).[1]

Yet, resource-based theory has evolved, for which new considerations include the role of the external business environment having an impact

[1] Firm performance can be measured (and assessed) in various capacities, including financial metrics, like sales and profits across different product-markets (Vorhies & Morgan, 2005; Young et al., 2000). Likewise, not all owner-managers are growth-oriented, and objectives may be lifestyle-oriented (Crick et al., 2016, 2018). This chapter views performance holistically, since owner-managers are likely to utilize an array of performance criteria (Crick & Crick, 2021b).

regarding influencing the performance outcomes of resources and capabilities (Priem & Butler, 2001). Additionally, this lens highlights that certain owner-managers require assistance from stakeholders (e.g., rivals) to achieve their day-to-day and longer-term objectives (Barney, 2018; Lavie, 2006). Hence, this investigation utilizes the wider elements of this theoretical lens in its conceptualizations.

Coopetition Strategies and Wine Tourism Offerings

Coopetition strategies can be implemented in multiple capacities across different product-markets (Bengtsson & Raza-Ullah, 2016; Bouncken et al., 2015; Crick & Crick, 2019; Gernsheimer et al., 2021; Meena et al., 2023).[2] Formal types of coopetition exist through organizations forming contractual alliances with their competitors, such as to enter export markets via pooling their resources and capabilities (Crick, 2021; Crick et al., 2024). Informal coopetition activities encapsulate non-contractual networks, like competing enterprises borrowing tools from their competitors or asking for advice from these key stakeholders on an ad hoc basis (Felzensztein et al., 2018; Granata et al., 2018).

Under resource-based theory, coopetition can assist firms to provide products and services in pursuit of opportunities that would be unlikely if they were to utilize only their own resources and capabilities (see Crick & Crick, 2021a). Regarding wine tourism offerings, by cooperating with rivals, vineyards and wineries can acquire stronger insights into catering to their end-users' wants and needs, together with acquiring tangible and intangible assets that can allow them to facilitate such offerings (Crick, 2018; Telfer, 2000). That is, wine tourism offerings are often costly and time-consuming to implement (Koch et al., 2013; Torres et al., 2021). Yet, by engaging in coopetition strategies, it follows

[2] Risks are involved in entering different product-markets in addition to opportunities, so owner-managers need to consider focusing on either one or a few markets or diversifying across various markets (Crick & Crick, 2023a; Crick et al., 2000). Therefore, decision-makers may use various stakeholder support to build capabilities, plus gain experience and information, other than through coopetition; for example, via government support providers (Chaudhry & Crick, 1998; Crick & Crick, 2016; Felzensztein et al., 2022; Jones & Crick, 2004; Sharma et al., 2018; Spence & Crick, 2001).

that entrepreneurs can facilitate wine tourism offerings that they perhaps would not engage in (or to a different extent) without the support of their competitors. Here, coopetition strategies could lower the costs of providing wine tourism offerings—by allowing a group of competing vineyards and wineries to join forces to supply value-adding products and services to new and existing customers (Crick & Crick, 2021c). With this in mind:

P1: Coopetition strategies facilitate wine tourism offerings.

Wine Tourism Offerings and Firm Performance

Like coopetition activities, wine tourism offerings manifest in various ways (Charters & Ali-Knight, 2002; Martinez-Falco et al., 2024). This includes anything from cellar door visits, vineyard/winery tours, through to more multi-faceted (and costly) products and services, such as weddings, accommodation, and partnerships with external organizations (e.g., golfing resorts) (Alant & Bruwer, 2004; Bruwer, 2003; Crick & Crick, 2015; Hall et al., 1997). Linking with resource-based theory, wine tourism offerings connect to the notion of a market orientation, namely, the organization-wide implementation of the marketing concept and creation of customer value (Hamzah et al., 2023; Hunt & Morgan, 1995; Jaworski & Kohli, 1993).

Wine tourism offerings are infused with a market orientation, since these firm-wide strategies are designed to create value for customers (Crick & Crick, 2021c; Rojas-Mendez & Rod, 2013). In other words, by supplying augmented products and services, vineyards and wineries can deliver value to their end-users by not just focusing on core wine sales, but also, providing an array of additional products and services that differentiate themselves from their rivals (Crick, 2015; Crick et al., 2020). Specifically, there could be a limit to the performance outcomes that a given wine producer could achieve by simply growing grapes and bottling wine. However, by providing augmented wine tourism offerings, entrepreneurs could achieve other performance-enhancing benefits, like attracting new customers, incentivizing existing customers, and beyond (Getz & Brown, 2006; Koch et al., 2013).

Such considerations are especially pertinent given that respective wine producers located in geographically close clusters[3] may know each other and have developed social capital (Crick & Crick, 2021d; Geldes et al., 2015). As such, owners may seek to enhance not only their own firm's performance (and maintain family employment where applicable), but also attempt to sustain their cluster via engaging in joint wine tourism activities (consistent with Reay et al., 2015). Thus, it is expected that:

P2: Wine tourism offerings lead to higher levels of firm performance.

Coopetition Strategies and Organizational Tensions

Consistent with resource-based theory, in all forms of coopetition (formal or informal), there are collaborative and rivalrous forces at play—no matter how small (Bengtsson & Kock, 2014; Crick & Crick, 2021d; Gnyawali et al., 2016; Hoffmann et al., 2018). Indeed, risk/reward considerations exist in selecting network partners as not all competitors are the same (Chiambaretto et al., 2020; Yan et al., 2020). Given the paradoxical nature of coopetition activities, there can be organizational tensions that emerge when competing enterprises join forces if not managed effectively (see, for example, Bouncken et al., 2018; Raza-Ullah et al., 2014). For instance, there can be diminishing levels of trust, increased conflict, elevated power imbalances, and so on, when competing firms share resources and capabilities (Bouncken & Kraus, 2013; Czakon & Czernek, 2016; Luo et al., 2007; Tidstrom, 2014).

In the wine industry, if vineyards and wineries supply wine tourism offerings (that are facilitated by coopetition activities), they might encounter challenges that are associated with being entangled in complex coopetition networks (building upon Crick et al., 2022a). For instance,

[3] Clusters are prominent in various sectors associated with hospitality and tourism (Chaudhry & Crick, 2004; Crick et al., 2016, 2018, 2023a; Karami et al., 2023), alongside others like agriculture and fisheries, alcohol, etc. (Felzensztein & Deans, 2013; Felzensztein et al., 2014, 2018).

jealousies or even opportunistic practices may occur when one coopetition partner succeeds in comparison to others in their network (Crick, 2015). Therefore, while there are advantages of engaging in coopetition strategies to facilitate wine tourism offerings, organizational tensions might emerge through ineffective management (i.e., potential dark-sides) (Crick, 2015). Consequently, it is anticipated that:

P3: Ineffectively managed coopetition strategies drive organizational tensions.

Organizational Tensions and Firm Performance

Grounded in the extended considerations of resource-based theory, organizational tensions (i.e., those driven through poorly managed coopetition networks) are problematic for businesses of different sizes (Bouncken et al., 2018; Raza-Ullah et al., 2014; Tidstrom, 2014). Specifically, by collaborating with unsuitable rivals, entrepreneurs (including wine producers) are likely to spend vast amounts of time and finances recovering from these problems, which in turn, restricts their efficiency and effectiveness—not least of which, struggling to deliver value to their customers (Crick & Crick, 2021a). Put another way, organizational tensions are a notable potential downside of coopetition strategies because they highlight that these business-to-business activities are paradoxical, as they are comprised of opposing (or perhaps complementary) cooperation and competition forces (Hoffmann et al., 2018; Mattsson & Tidstrom, 2015). If organizational tensions emerge through ineffectively managed practices, it is likely that vineyards and wineries will experience performance-reducing consequences, like lost customers and market share (consistent with Crick et al., 2022a). Consequently, it stands to reason that:

P4: Organizational tensions reduce firm performance.

Coopetition Strategies and Firm Performance

To connect the previous aspects of the conceptual model, while there may be associations linking coopetition strategies, wine tourism offerings, and firm performance, there may be a direct relationship between coopetition strategies and firm performance (in line with Crick, 2022; Ritala, 2012). As previously mentioned, coopetition activities can assist entrepreneurs to deliver value to their customers in new and improved ways (Bengtsson & Kock, 2000; Crick & Crick, 2023b). For example, in the capacity of the COVID-19 pandemic, Crick and Crick (2020) found that organizations of different industries and sizes pivoted their business models in response to the health and economic disasters that were associated with this large-scale emergency. As an illustration, Crick and Crick (2020) identified that retailers in the United Kingdom shared information about inventories, coordinated their opening/closing hours, and loaned employees to rival stores—versus operating on a purely rivalrous basis.

Hence, there is evidence to suggest that if firms (small and large—including vineyards and wineries) effectively collaborate with their competitors, they can acquire new ways to survive and grow within their markets (Crick, 2015; Crick & Crick, 2022). This encapsulates wine producers organizing events, borrowing tools, equipment, and asking for knowledge from rivals that elevate their efficiency and effectiveness at creating value for their end-users (following Felzensztein & Deans, 2013; Granata et al., 2018; Telfer, 2000). Therefore, it follows that:

P5a: Coopetition strategies have a linear link with firm performance.

Nevertheless, P5a may be somewhat simplistic in capturing the relationship between coopetition and performance. There is a body of research indicating that although there are key advantages of coopetition strategies, there are potential dark-sides of these business-to-business marketing activities (Crick et al., 2022a; Luo et al., 2007; Park et al., 2014). In the capacity of the wine industry, Crick (2020) found that in New Zealand, vineyards and wineries can obtain performance-enhancing benefits of coopetition activities, such as access to new tangible and intangible assets, as well as opportunities. However, these benefits are sometimes superseded by the potential dark-sides of the business-to-business networks, like the earlier-mentioned organizational tensions (consistent with Bouncken et al., 2018; Mattsson & Tidstrom, 2015). This means that instead of there being a linear (and positive) association between coopetition strategies and firm performance, this connection is more likely to be non-linear (inverted U-shaped) (in line with Crick & Crick, 2021b; Crick et al., 2024). Accordingly, and as an alternate viewpoint to P5a, it is proposed that:

P5b: Coopetition strategies have a non-linear (inverted U-shaped) link with firm performance.

Even so, there may be contingencies (through moderating factors) that influence the connection between coopetition strategies and firm performance (Crick & Crick, 2021e; Hoffmann et al., 2018). Therefore, consistent with wider tourism/hospitality research, it is important that decision-makers learn what approaches work and do not work, and pivot facets of their business model (Crick, 2011). Regarding the external business environment, there could be market-level forces that impact the extent to which coopetition can elevate firm performance (in line with Hoffmann et al., 2018; Park et al., 2014; Ritala, 2012). For instance, Crick and Crick (2021f) found that wine producers in New Zealand might experience a diminishing-returns effect when cooperating with their competitors—a function of a non-linear (inverted U-shaped) link between coopetition strategies and firm performance. Yet, Crick and Crick (2021f) added that competitive intensity could amplify this quadratic association, in which rivalrous market-wide conditions propel the potential dark-sides of behavioral forms of coopetition (e.g., organizational tensions). Hence, it stands to reason that the decision-makers within vineyards and wineries might struggle to achieve the performance-enhancing benefits of coopetition activities in intensely competitive markets. Thus, it is argued that:

P6: Competitive intensity negatively influences the potentially quadratic link between coopetition strategies and firm performance.

Conceptual Model

Under the extended dimensions of resource-based theory (Barney, 2018; Lavie, 2006; Priem & Butler, 2001), this chapter unpacks the intricacies of coopetition strategies in the capacity of wine tourism providers. The conceptual model was comprised of research propositions that were used to achieve the study's research objective (Fig. 4.1). Some practitioner implications are offered in the next section.

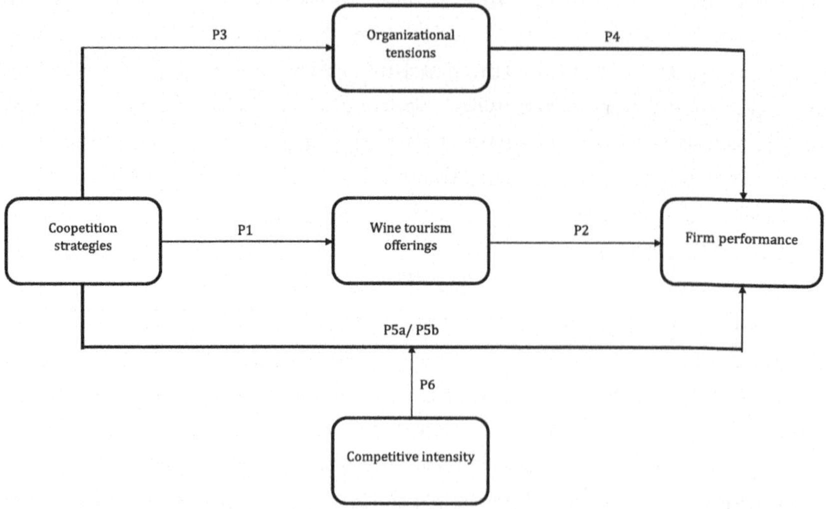

Fig. 4.1 Conceptual model and research propositions

4.3 Practitioner Implications

Alongside its contributions to the extant literature, this chapter offers the following recommendations for practitioners (focusing on entrepreneurs within wine-producing countries):

1. Cooperating with competitors can be a beneficial business-to-business activity, as formal and informal types of coopetition can assist entrepreneurs to obtain new assets like funds, equipment, and knowledge that allows them to survive and grow within their markets. Consequently, entrepreneurs are recommended to explore opportunities for engaging in coopetition strategies to help boost their market-level successes.
2. Decision-makers should nonetheless be conscious of the potential dark-sides of cooperating with their rivals. These problems encapsulate organizational tensions (e.g., conflict, power imbalances, and diminished trust) that could occur when rival enterprises work together, which can harm firm performance due to the costs (financial or otherwise) that can accumulate when these business-to-business activities are mismanaged. For example, problems can occur when engaging with untrustworthy or non-complementary rivals. Additionally, when there is no exit strategy envisaged if things do not work out as planned.
3. Coopetition strategies can be particularly problematic when sectors (or wine regions) are fiercely competitive. To that end, vineyards and wineries are likely to experience the potential dark-sides of coopetition strategies in intensely competitive markets. This offers the prospect of highlighting a situation where coopetition may (or may not) be a performance-enhancing business-to-business activity; hence, requiring effective management.
4. Coopetition activities can assist wine producers to facilitate wine tourism offerings. That is, by sharing resources and capabilities with trustworthy and complementary competing enterprises, vineyards and wineries can implement augmented products and services that supplement their core offerings (i.e., bottles of wine). In turn, these

business-to-business networks can overcome decision-makers' limited resources and capabilities.
5. If wine producers offer augmented products and services (via wine tourism offerings like restaurants, accommodation, etc.), they can use them to deliver value to their customers. In fact, coopetition can lead to cross-referrals between network partners with varying wine tourism offerings. As such, wine tourism offerings can assist wine producers to boost their market-level successes—over and above traditional products and services (like core bottles of wine).
6. If vineyards and wineries engage in coopetition, they can promote their regions domestically and internationally—leading to potentially more sales and profits, as well as increased reputations with customers in different parts of the world. This is especially true if regions are relatively "unheard of", meaning that coopetition can facilitate opportunities across regional and national borders. That said, as mentioned above, there are potential risks when coopetition strategies are implemented, so it is of paramount importance that caution is exercised in these networks.

Collectively, these practitioner implications are designed to assist wine producers to increase firm performance when engaging in coopetition strategies, coupled with supplying wine tourism offerings. A series of conclusions and future research directions follow in the next section.

4.4 Conclusions and Future Research Directions

Under the wider themes of resource-based theory (following Barney, 2018; Lavie, 2006), the objective of this chapter was to examine how coopetition strategies can be used to facilitate wine tourism strategies. This objective was achieved through reviewing the extended coopetition literature, together with earlier work that pertains to wine tourism offerings. This study developed a conceptual model (with various research propositions) covering the intricacies of the linkages between coopetition strategies, wine tourism offerings, and firm performance. In doing so, it is concluded that while coopetition strategies can assist wine producers to boost their market-level successes, there are potential problems related to these inter-firm networks, such as organizational tensions (extending Crick, 2020; Mattsson & Tidstrom, 2015; Raza-Ullah et al., 2014; Tidstrom, 2014).

Another conclusion is that coopetition strategies can facilitate wine tourism offerings, which can be used to help wine producers to boost firm performance through market-oriented behaviors that can create value for end-users (linking with Crick, 2018; Rojas-Mendez & Rod, 2013; Telfer, 2000; Torres et al., 2021). It is also concluded that competitive intensity has scope to negatively influence the potentially quadratic connection between coopetition strategies and firm performance (extending Crick & Crick, 2021b; Hoffmann et al., 2018; Park et al., 2014; Ritala, 2012). Finally, this chapter has shed new light on the wider elements of resource-based theory, with respect to how cooperating with competitors assists enterprises (of different sizes) to cope with the ongoing pressures within their sectors—not least of which competitively intense markets (building upon Crick & Crick, 2021f; Lavie, 2006).

Moving forward, this chapter has developed the existing body of knowledge pertaining to the connections between coopetition strategies, wine tourism offerings, and firm performance. However, this merely serves as a starting point to commence new lines-of-thinking regarding these issues and themes. In other words, there are a series of future research directions pertaining to ongoing research concerning vineyards and wineries in the years to come. First, future research should address the explicit advantages and disadvantages of coopetition activities, in terms of the situations where cooperating with competitors is (and is not) beneficial for vineyards and wineries (following Crick et al., 2024; Dana et al., 2013; Felzensztein et al., 2014; Granata et al., 2018; Telfer, 2000). Second, academics are advised to unpack how wine tourism offerings can be implemented. For example, how entrepreneurs can deliver value to their customers through augmented products and services—and avoiding being reliant on supplying core bottles of wine to their end-users (developing Alant & Bruwer, 2004; Bruwer, 2003; Charters & Ali-Knight, 2002; Getz & Brown, 2006; Hall et al., 1997; Martinez-Falco et al., 2024).

Third, it would be interesting to enhance the use of resource-based theory as a lens that can be utilized to unpack firm-wide behaviors pertaining to different wine-producing countries (consistent with Crick, 2015; Crick & Crick, 2023b). Research might empirically assess the connection between coopetition activities, wine tourism offerings, and firm performance in several wine regions and countries. This includes obtaining insights from the Old World and New World wine-producing countries to explore if differences exist between organizations (small and large) that engage in wine tourism strategies, alongside joining forces with rival enterprises to increase their market-level successes (building upon Crick & Crick, 2015; Crick et al., 2024; Granata et al., 2018; Telfer, 2000; Torres et al., 2021). It would be especially helpful to operationalize and test these proposed constructs and relationships with statistical data to yield generalizable results. To summarize, these directions for future research are intended to enhance the extant literature by taking the themes of this chapter—and use them to conduct novel future conceptual and empirical research.

References

Alant, K., & Bruwer, J. (2004). Wine tourism behaviour in the context of a motivational framework for wine regions and cellar doors. *Journal of Wine Research, 15*(1), 27–37.

Barney, J. B. (1991). Firm resources and sustained competitive advantage. *Journal of Management, 17*(1), 99–120.

Barney, J. B. (2018). Why resource-based theory's model of profit appropriation must incorporate a stakeholder perspective. *Strategic Management Journal, 39*(13), 3305–3325.

Bengtsson, M., & Kock, S. (1999). Cooperation and competition in relationships between competitors in business networks. *Journal of Business & Industrial Marketing, 14*(3), 178–194.

Bengtsson, M., & Kock, S. (2000). Coopetition in business networks: To cooperate and compete simultaneously. *Industrial Marketing Management, 29*(5), 411–426.

Bengtsson, M., & Kock, S. (2014). Coopetition—Quo vadis? Past accomplishments and future challenges. *Industrial Marketing Management, 43*(2), 180–188.

Bengtsson, M., & Raza-Ullah, T. (2016). A systematic review of research on coopetition: Toward a multi-level understanding. *Industrial Marketing Management, 57*(1), 23–39.

Bengtsson, M., Raza-Ullah, T., & Vanyushyn, V. (2016). The coopetition paradox and tension: The moderating role of coopetition capability. *Industrial Marketing Management, 53*(1), 19–30.

Bouncken, R. B., & Fredrich, V. (2012). Coopetition: Performance implications and management antecedents. *International Journal of Innovation Management, 16*(5), 1250028.

Bouncken, R. B., Fredrich, V., Ritala, P., & Kraus, S. (2018). Coopetition in new product development alliances: Advantages and tensions for incremental and radical innovation. *British Journal of Management, 29*(3), 391–410.

Bouncken, R. B., Fredrich, V., Ritala, P., & Kraus, S. (2020, October). Value-creation-capture-equilibrium in new product development alliances: A matter of coopetition, expert power, and alliance importance. *Industrial Marketing Management, 90*, 648–662.

Bouncken, R. B., Gast, J., Kraus, S., & Bogers, M. (2015). Coopetition: A systematic review, synthesis, and future research directions. *Review of Managerial Science, 9*(3), 577–601.

Bouncken, R. B., & Kraus, S. (2013). Innovation in knowledge-intensive industries: The double-edged sword of coopetition. *Journal of Business Research, 66*(10), 2060–2070.

Brandenburger, A. M., & Nalebuff, B. J. (1996). *Co-opetition.* Doubleday Dell Publishing Group Inc.

Bruwer, J. (2003). South African wine routes: Some perspectives on the wine tourism industry's structural dimensions and wine tourism product. *Tourism Management, 24*(4), 423–435.

Charters, S., & Ali-Knight, J. (2002). Who is the wine tourist? *Tourism Management, 23*(3), 311–319.

Chaudhry, S., & Crick, D. (1998). Export information providers: Are they meeting the needs of SMEs? *Marketing Intelligence and Planning, 16*(3), 141–149.

Chaudhry, S., & Crick, D. (2004). The business practices of small Chinese restaurants in the U.K: An exploratory investigation. *Journal of Strategic Change, 13*(1), 37–49.

Chiambaretto, P., Bengtsson, M., Fernandez, A.-S., & Näsholm, N. H. (2020). Small and large firms' trade-off between benefits and risks when choosing a coopetitor for innovation. *Long Range Planning, 53*(1), 101876.

Crick, D. (2011). Enterprising individuals and entrepreneurial learning: A longitudinal case history in the UK tourism sector. *International Journal of Entrepreneurial Behaviour & Research, 17*(2), 203–218.

Crick, D., Chaudhry, S., & Batstone, S. (2000). Revisiting the concentration versus spreading debate as a successful growth strategy: The case of UK SMEs exporting agricultural related products. *Entrepreneurship and Regional Development, 12*(1), 49–67.

Crick, D., Chaudhry, S., & Crick, J. M. (2016). Trading in a competitive environment: South-Asian restaurants in the UK. *Strategic Change: Briefings in Entrepreneurial Finance, 25*(4), 371–382.

Crick, D., Chaudhry, S., & Crick, J. M. (2018). Risks/rewards and an evolving business model: A case study of a small lifestyle business in the UK tourism sector. *Qualitative Market Research: An International Journal, 21*(2), 143–165.

Crick, D., & Crick, J. M. (2015). Learning and decision-making in marketing planning: A study of New Zealand vineyards. *Marketing Intelligence & Planning, 33*(5), 707–732.

Crick, D., & Crick, J. M. (2016). An appreciative inquiry into the first export order. *Qualitative Market Research: An International Journal, 19*(1), 84–100.

Crick, J. M. (2015). *Bridging the gap between threshold and dynamic capabilities: A qualitative study of the collaboration strategies of New Zealand wineries* (Unpublished MBS thesis in Management). Massey University, New Zealand.

Crick, J. M. (2018). Studying coopetition in a wine industry context: Directions for future research. *International Journal of Wine Business Research, 30*(3), 366–371.

Crick, J. M. (2019). Incorporating coopetition into the entrepreneurial marketing literature: Directions for future research. *Journal of Research in Marketing and Entrepreneurship, 21*(1), 19–36.

Crick, J. M. (2020). The dark-side of coopetition: When collaborating with competitors is harmful for company performance. *Journal of Business & Industrial Marketing, 35*(2), 318–337.

Crick, J. M. (2021). Unpacking the relationship between a coopetition-oriented mind-set and coopetition-oriented behaviours. *Journal of Business & Industrial Marketing, 36*(3), 400–419.

Crick, J. M. (2022). Does competitive aggressiveness negatively moderate the relationship between coopetition and customer satisfaction performance? *Journal of Strategic Marketing, 30*(6), 562–587.

Crick, J. M., & Crick, D. (2019). Developing and validating a multi-dimensional measure of coopetition. *Journal of Business & Industrial Marketing, 34*(4), 665–689.

Crick, J. M., & Crick, D. (2020). Coopetition and COVID-19: Collaborative business-to-business marketing strategies in a pandemic crisis. *Industrial Marketing Management, 88*(1), 206–213.

Crick, J. M., & Crick, D. (2021a). Coopetition and family-owned wine producers. *Journal of Business Research, 135*(1), 319–336.

Crick, J. M., & Crick, D. (2021b). Internationalizing the coopetition construct: Quadratic effects on financial performance under different degrees of export intensity and an export geographical scope. *Journal of International Marketing, 29*(2), 62–80.

Crick, J. M., & Crick, D. (2021c, July). Rising up to the challenge of our rivals: Unpacking the drivers and outcomes of coopetition activities. *Industrial Marketing Management, 96*, 71–85.

Crick, J. M., & Crick, D. (2021d). The dark-side of coopetition: Influences on the paradoxical forces of cooperativeness and competitiveness across product-market strategies. *Journal of Business Research, 122*(1), 226–240.

Crick, J. M., & Crick, D. (2021e). Coopetition and sales performance: Evidence from non-mainstream sporting clubs. *International Journal of Entrepreneurial Behavior & Research, 27*(1), 123–147.

Crick, J. M., & Crick, D. (2021f). The yin and yang nature of coopetition activities: Non-linear effects and the moderating role of competitive intensity for internationalised firms. *International Marketing Review, 38*(4), 690–716.

Crick, J. M., & Crick, D. (2022). Coopetition and international entrepreneurship: The influence of a competitor orientation. *International Journal of Entrepreneurial Behavior & Research, 28*(3), 801–828.

Crick, J. M., & Crick, D. (2023a). Revisiting the 'concentration vs spreading debate': Perceived risk and strategic flexibility in decision-making regarding an unanticipated environmental market disruption. *Journal of Strategic Marketing, 31*(3), 578–606.

Crick, J. M., & Crick, D. (2023b). With a little help from my friends: The interaction between coopetition, an entrepreneurial orientation, and firm performance. *International Journal of Entrepreneurial Behavior & Research, 29*(4), 965–985.

Crick, J. M., & Crick, D. (2024, forthcoming). Regional-level coopetition strategies and company performance: Evidence from the Canadian wine industry. *Entrepreneurship & Regional Development: An International Journal*.
Crick, J. M., Crick, D., & Chaudhry, S. (2022a). The dark-side of coopetition: It's not what you say, but the way that you do it. *Journal of Strategic Marketing, 30*(1), 22–44.
Crick, J. M., Crick, D., & Chaudhry, S. (2023a). Inter-firm collaboration as a performance-enhancing survival strategy within the business models of ethnic minority-owned urban restaurants affected by COVID-19. *International Journal of Entrepreneurial Behavior & Research, 29*(3), 587–613.
Crick, J. M., Crick, D., & Peixinho, J. (2023b). Does industry experience positively moderate the quadratic relationship between coopetition and financial performance? Evidence from the New Zealand wine sector. *International Journal of Entrepreneurship and Small Business, 49*(1), 1–31.
Crick, J. M., Crick, D., & Tebbett, N. (2020). Competitor orientation and value co-creation in sustaining rural New Zealand wine producers. *Journal of Rural Studies, 73*(1), 122–134.
Crick, J. M., Karami, M., & Crick, D. (2021). The impact of the interaction between an entrepreneurial marketing orientation and coopetition on business performance. *International Journal of Entrepreneurial Behavior & Research, 27*(6), 1423–1447.
Crick, J. M., Karami, M., & Crick, D. (2022b, January). Is it enough to be market-oriented? How coopetition and industry experience affect the relationship between a market orientation and customer satisfaction performance. *Industrial Marketing Management, 122*, 62–75.
Crick, J. M., Crick, D., & Ferrigno, G. (2024, forthcoming). Coopetition and the marketing/entrepreneurship interface in an international arena. *International Journal of Entrepreneurial Behavior & Research*.
Czakon, W., & Czernek, K. (2016). The role of trust-building mechanisms in entering into network coopetition: The case of tourism networks in Poland. *Industrial Marketing Management, 57*(1), 64–74.
Czakon, W., Klimas, P., & Mariani, M. (2020a). Behavioral antecedents of coopetition: A synthesis and measurement scale. *Long Range Planning, 53*(1), 101875.
Czakon, W., Srivastava, M. K., Le Roy, F., & Gnyawali, D. (2020b). Coopetition strategies: Critical issues and research directions. *Long Range Planning, 53*(1), 101948.

Dana, L.-P., Granata, J., Lasch, F., & Carnaby, A. (2013). The evolution of co-opetition in the Waipara wine cluster of New Zealand. *Wine Economics and Policy, 2*(1), 42–49.

Estrada, I., & Dong, J. Q. (2020). Learning from experience? Technological investments and the impact of coopetition experience on firm profitability. *Long Range Planning, 53*(1), 101866.

Felzensztein, C., Crick, D., Gonzalez-Perez, M. A., Jurado, T., & Etchebarne Lopez, M. S. (2022). Capabilities and the internationalisation of smaller-sized, service-oriented firms in the southern hemisphere. *Journal of Strategic Marketing, 30*(6), 533–561.

Felzensztein, C., & Deans, K. R. (2013). Marketing practices in wine clusters: Insights from Chile. *Journal of Business & Industrial Marketing, 28*(4), 357–367.

Felzensztein, C., Gimmon, E., & Deans, K. R. (2018). Coopetition in regional clusters: Keep calm and expect unexpected changes. *Industrial Marketing Management, 69*(1), 116–124.

Felzensztein, C., Stringer, C., Benson-Rea, M., & Freeman, S. (2014). International marketing strategies in industrial clusters: Insights from the Southern Hemisphere. *Journal of Business Research, 67*(5), 837–846.

Geldes, C., Felzensztein, C., Turkina, E., & Durand, A. (2015). How does proximity affect inter-firm marketing cooperation? A study of an agribusiness cluster. *Journal of Business Research, 68*(2), 263–272.

Gernsheimer, O., Kanbach, D. K., & Gast, J. (2021, July). Coopetition research—A systematic literature review on recent accomplishments and trajectories. *Industrial Marketing Management, 96*, 113–134.

Getz, D., & Brown, G. (2006). Critical success factors for wine tourism regions: A demand analysis. *Tourism Management, 27*(1), 146–158.

Gnyawali, D. R., Madhavan, R., He, J., & Bengtsson, M. (2016). The competition–cooperation paradox in inter-firm relationships: A conceptual framework. *Industrial Marketing Management, 53*(1), 7–18.

Granata, J., Lasch, F., Le Roy, F., & Dana, L.-P. (2018). How do micro-firms manage coopetition? A study of the wine sector in France. *International Small Business Journal: Researching Entrepreneurship, 36*(3), 331–355.

Hall, C. M., Cambourne, B., Macionis, N., & Johnson, G. (1997). Wine tourism and network development in Australia and New Zealand: Review, establishment, and prospects. *International Journal of Wine Marketing, 9*(2), 5–31.

Hamzah, M. I., Crick, J. M., Crick, D., Ali, S. A. M., & Yunus, N. M. (2023). The nature of the relationship between an entrepreneurial marketing orientation and small business growth: Evidence from Malaysia. *International Journal of Entrepreneurship and Small Business, 50*(3), 355–391.

Hoffmann, W., Lavie, D., Reuer, J. J., & Shipilov, A. (2018). The interplay of competition and cooperation. *Strategic Management Journal, 39*(12), 3033–3052.

Hunt, S. D., & Morgan, R. M. (1995). The comparative advantage theory of competition. *Journal of Marketing, 59*(2), 1–15.

Jaworski, B. J., & Kohli, A. K. (1993). Market orientation: Antecedents and consequences. *Journal of Marketing, 57*(3), 53–70.

Jones, M. V., & Crick, D. (2004). Internationalising high-technology-based UK firms' information-gathering activities. *Journal of Small Business and Enterprise Development, 11*(1), 84–94.

Karami, M., Crick, D., & Crick, J. M. (2023). Non-predictive decision-making, market-oriented behaviours, and smaller-sized firms' performance. *Journal of Strategic Marketing, 31*(5), 1107–1131.

Koch, J., Martin, A., & Nash, R. (2013). Overview of perceptions of German wine tourism from the winery perspective. *International Journal of Wine Business Research, 25*(1), 50–74.

Kraus, S., Klimas, P., Gast, J., & Stephan, T. (2019). Sleeping with competitors: Forms, antecedents and outcomes of coopetition of small and medium-sized craft beer breweries. *International Journal of Entrepreneurial Behavior & Research, 25*(1), 50–66.

Lavie, D. (2006). The competitive advantage of inter-connected firms: An extension of the resource-based view. *Academy of Management Review, 31*(3), 638–658.

Luo, X., Rindfleisch, A., & Tse, D. K. (2007). Working with rivals: The impact of competitor alliances on financial performance. *Journal of Marketing Research, 44*(1), 73–83.

Martinez-Falco, J., Marco-Lajara, B., Zaragoza-Saez, P. D. C., & Millan-Tudela, L. A. (2024). Do circular economy practices moderate the wine tourism—Green performance relationship? A structural analysis applied to the Spanish wine industry. *British Food Journal, 126*(1), 134–155.

Mattsson, L.-G., & Tidstrom, A. (2015). Applying the principles of yin–yang to market dynamics: On the duality of cooperation and competition. *Marketing Theory, 15*(3), 347–364.

Meena, A., Dhir, S., & Sushil, S. (2023). A review of coopetition and future research agenda. *Journal of Business & Industrial Marketing, 38*(1), 118–136.

Park, B. J. R., Srivastava, M. K., & Gnyawali, D. R. (2014). Walking the tight rope of coopetition: Impact of competition and cooperation intensities and balance on firm innovation performance. *Industrial Marketing Management, 43*(2), 210–221.

Priem, R. L., & Butler, J. E. (2001). Is the resource-based view a useful perspective for strategic management research? *Academy of Management Review, 26*(1), 57–66.

Raza-Ullah, T., Bengtsson, M., & Kock, S. (2014). The coopetition paradox and tension in coopetition at multiple-levels. *Industrial Marketing Management, 43*(2), 189–198.

Reay, T., Jaskiewicz, P., & Hinings, C. R. (2015). How family, business, and community logics shape family firm behaviour and "rules of the game" in an organizational field. *Family Business Review, 28*(4), 292–311.

Ritala, P. (2012). Coopetition strategy—When is it successful? Empirical evidence on innovation and market performance. *British Journal of Management, 23*(3), 307–324.

Rojas-Mendez, J. I., & Rod, M. (2013). Chilean wine producer market orientation: Comparing MKTOR versus MARKOR. *International Journal of Wine Business Research, 25*(1), 27–49.

Sharma, R. R., Sraha, G., & Crick, D. (2018). Export promotion programmes and the export performance of Ghanaian firms: The mediating role of foreign market attractiveness. *International Marketing Review, 35*(4), 661–682.

Spence, M. M., & Crick, D. (2001). An investigation into UK firms' use of trade missions. *Marketing Intelligence & Planning, 19*(7), 464–474.

Telfer, D. J. (2000). The North East wine route. In C. M. Hall, L. Sharples, B. Cambourne, & N. Macionis (Eds.), *Tourism around the world* (pp. 1–11). Butterworth Heinemann.

Tidstrom, A. (2014). Managing tensions in coopetition. *Industrial Marketing Management, 43*(2), 261–271.

Torres, J. P., Barrera, J. I., Kunc, M., & Charters, S. (2021). The dynamics of wine tourism adoption in Chile. *Journal of Business Research, 127*(1), 474–485.

Vorhies, D. W., & Morgan, N. A. (2005). Benchmarking marketing capabilities for sustainable competitive advantage. *Journal of Marketing, 69*(1), 80–94.

Yan, Y., Dong, J. Q., & Faems, D. (2020). Not every coopetitor is the same: The impact of technological, market and geographical overlap with coopetitors on firms' breakthrough inventions. *Long Range Planning, 53*(1), 101873.

Young, S., Bell, J., & Crick, D. (2000). The resource-based perspective and small firm internationalisation: An exploratory approach. In C. C. J. M. Millar, R. M. Grant, & C. J. Choi (Eds.), *International business*. The Academy of International Business Series (pp. 79–101). Palgrave Macmillan.

Part III

Demand and Consumer Behavior in Wine Tourism Sector

5

Message in a Bottle: Generation Y and Wine Purchase Occasions' Perceived Risks and Information Sources

Ana Brochado, Helena Rodrigues, and Sara Vaz Mendes

5.1 Introduction

The wine market is saturated. Consumers frequently become confused when selecting wine due to the diverse range of brands, prices and styles (Corduas et al., 2013). Much single-market and cross-national research has been done on consumers' wine behaviour to understand

A. Brochado
Centre for Socioeconomic and Territorial Studies (DINÂMIA'CET)-IUL, Lisbon, Portugal
e-mail: Ana.Brochado@iscte-iul.pt

H. Rodrigues (✉)
Instituto Superior de Ciências do Trabalho e da Empresa (ISCTE)-Instituto Universitário de Lisboa (IUL) and Business Research Unit-IUL, Lisbon, Portugal
e-mail: hmcnc@iscte-iul.pt

S. V. Mendes
Instituto Superior de Ciências do Trabalho e da Empresa (ISCTE)-Instituto Universitário de Lisboa (IUL), Lisbon, Portugal

J. Martínez-Falcó et al. (eds.), *Strategic Management in the Wine Tourism Industry*,
https://doi.org/10.1007/978-3-031-54837-6_5

specific wine markets' unique characteristics or to compare different wine markets (Lockshin & Corsi, 2012). Customers tend to purchase bottles of wine for specific occasions (Aqueveque, 2023; Boncinelli et al., 2019), so various authors have focused on distinguishing between types of consumption situations. These scholars have observed that the list of circumstances influencing wine purchasing behaviour needs to be expanded and developed more fully (Bruwer et al., 2013; Sandell, 1968). Studies in different countries are especially crucial as the findings could be more valuable to specific market players.

Segmentation based on specific criteria helps marketers identify needs and thus allocate their budget better to strengthen more effective marketing strategies (Barber et al., 2008). The literature highlights that different occasions can influence final wine purchase and wine consumption decisions, so this topic requires further research. Related traditions usually include festive occasions and gifts (Carsana & Jolibert, 2017), as well as individuals' personal use of wine (Boncinelli et al., 2019). Most often, exceptional occasions cause consumers to spend time and money shopping for high-quality wine (Bruwer et al., 2013). The study discussed in this chapter thus examined wine consumers' behaviour in different wine situations to answer previous calls for research on a more diverse set of wine consumption situations.

5.2 Literature Review

Millennials' Wine Consumption

Because culture is a significant determinant of consumption, wine-drinking behaviour differs across countries and occasions (Fountain & Lamb, 2011). Hall et al. (1997) conducted a cross-cultural investigation of wine consumers in Australia and Europe, focusing on their wine intake on different occasions and revealing major divergences that may be due to contrasting cultural backgrounds. Other studies have found that specific occasions worldwide influence wine purchase behaviour, but their implications are perceived differently depending on the culture involved (Carsana & Jolibert, 2017).

In addition, Generation Y's wine-drinking habits differ greatly from those of earlier generations (Williamson et al., 2016). This age group is building a wine-related culture in the United States and Australia that was previously non-existent, contrary to what can be seen in France and Italy. Espejel et al. (2011) found that Generation Y drinks more beer and less wine than previous generations. Millennials appear to be concerned about healthier and more responsible drinking, so the wine industry has introduced sustainable practices in, among other things, wine labelling (Sogari et al., 2015). Due to pressure from these consumers, wine producers' current main goal is to stop using herbicides and pesticides in vineyards.

Wine enthusiasts, in general, look for information from wine-related communities (Thach, 2009). According to Fernández et al. (2017), marketers are thus spending more on social media, blogs and even wine applications (Beninger et al., 2014).

Portuguese Millennials' Wine Consumption Behaviour

Portugal has one of the strongest wine traditions worldwide (Brochado et al., 2021, 2022). However, academics have not yet studied market segmentation in the Portuguese wine market more closely (Brochado & Martins, 2008; Figueiredo et al., 2003), especially regarding Generation Y's behaviour. Wine consumption varies within each country because of regional and generational cultural patterns, which are crucial factors shaping attitudes and conduct related to wine on different occasions (Fountain & Lamb, 2011).

Males between 35 and 49 years old are the leading wine drinkers among Portuguese consumers (Duarte et al., 2010). Low-cost wines are frequently bought at retail establishments, so the cited authors also concluded that price is an important determining factor of wine purchasing behaviour. Duarte et al. (2010) nonetheless found that, depending on the occasion, Generation X males are willing to spend more for a bottle of wine (Ferreira et al., 2019).

In addition, a small number of individuals under 24 years old prefer to consume wine in restaurants, but most Portuguese say that they would rather drink wine at home (Figueiredo et al., 2003). Fifty-two per cent of the latter consumers report that they offer wine on special occasions when dining with friends. According to Sanches (2013), Portuguese consumers rank, by order of importance, of the spectrum off-premises sales namely in hypermarkets and supermarkets, and on-premises such as restaurants as places to buy wine. Figueiredo et al. (2003) found that off-premises sales in these establishments are due to their strategic location and prices are key factors in why Generation Y buy wine at hypermarkets and supermarkets. Furthermore, wine clubs have a minimal expression in the Portuguese market, and winery websites are mostly used for formation searches rather than online purchases; this might be explained by the young demographic group that is part of this study (Figueiredo et al., 2003; Sellers & Nicolau, 2023). Another interesting conclusion was that Portuguese consumers place a high value on direct purchases from producers; nevertheless, those places are not always easily accessible (e.g. they have reduced visiting hours and are difficult to access).

Wine Occasions

Wines are currently marketed for every event and season, as well as pairings with all types of food. Wine occasions have been studied for many decades. Belk (1974) was one of the first researchers to confirm that customers choose different wines for specific occasions. Nicholson subsequently extended Belk's (1974) work with a study conducted in the United Kingdom, which identified the situations in which customers choose to drink wine as (1) with meals during special occasions, (2) while dining out or (3) when cooking.

Bruwer et al. (2013) in turn combined Quester and Smart (1998) and Hall et al.'s (2001) occasion models to create a more detailed list: occasions that are at home, intimate, with family, with friends, business and gift-giving and/or celebrations (see Fig. 5.1). The cited research verified

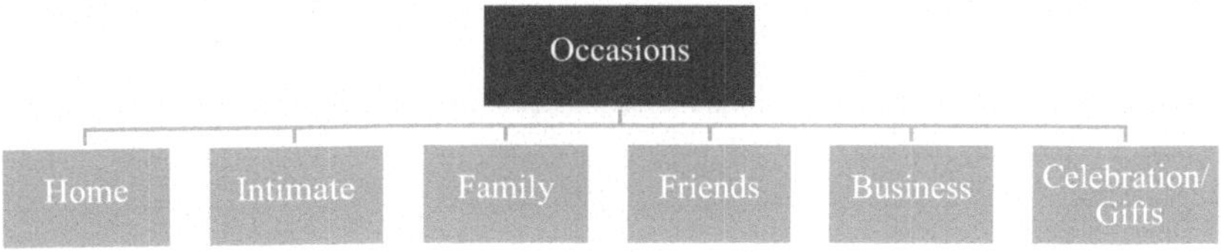

Fig. 5.1 Multidirectional conceptual framework—occasions (*Source* Adapted from Bruwer et al. [2013])

that friends, family members, partners and colleagues are most often present at wine consumption occasions, which gives these individuals considerable influence over wine consumers' choices.

Wine Attributes' Influence

Corduas et al. (2013) and Marques and Guia (2015) list the wine attributes that affect purchase choices as colour, grape variety, brand, labelling, alcohol percentage, region and country of origin, among others. Boncinelli et al. (2019) narrowed the key attributes down to only four: price, geographical indication, organic claim and brand. Prior studies have not considered label characteristics, and various studies have shown that, when consumers are questioned directly about wine attributes, labels are seen as the least essential feature (Brochado & Martins, 2008; Goodman, 2009; Mueller et al., 2011; Orth & Malkewitz, 2008). However, the label becomes significant during purchases in which value is measured indirectly because labels communicate the full range of extrinsic wine characteristics (Mueller et al., 2011) (see Fig. 5.2). Wine producers are aware of consumer preferences and prepare a wide range of wines to match these predilections (Goodman, 2009).

Perceived Risks' Influence

The literature shows that perceived risks have not been investigated for wine consumers (Bruwer & Cohen, 2017; Bruwer et al., 2009). Wine purchases imply an evaluation of a number of perceived risks, and

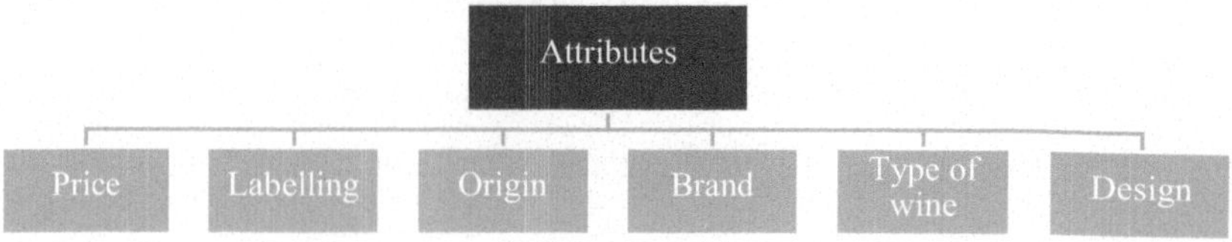

Fig. 5.2 Multidirectional conceptual framework—attributes (*Source* Authors)

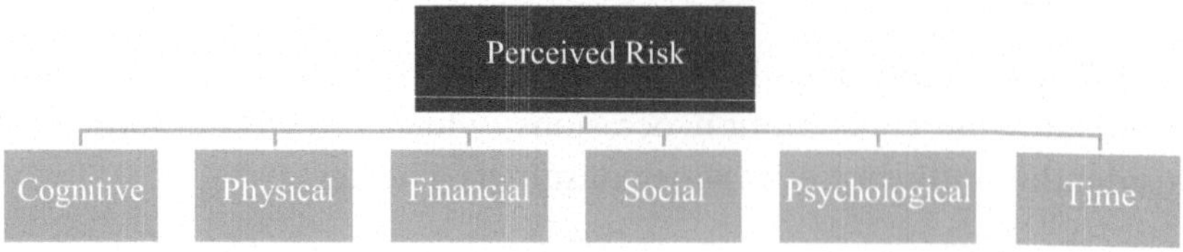

Fig. 5.3 Directional conceptual framework—perceived risks (*Source* Adapted from Bruwer et al. [2013])

consumers usually pick products based on which appear to be less risky (Figueiredo et al., 2003). Jacoby and Kaplan (1972) identified five types of risks: (1) cognitive, (2) social, (3) financial, (4) psychological and (5) physical. Schiffman et al. (2011) added time risks, thereby creating a model of six generic risk categories that Bruwer et al. (2013) used to develop a measurement scale of perceived risk in wine purchase decisions for specific occasions (see Fig. 5.3).

Information Sources' Influence

Consumers search for information to alleviate their fear of perceived risks (Bruwer et al., 2013; Dowling & Staelin, 1994; Murray, 1991). Wine producers tend to employ risk-reducing measures based on prospective consumers' frequent search for information from multiple sources to help them deal with uncertain situations or make purchase decisions (Day, 1970; Locander & Hermann, 1979). This behaviour is particularly

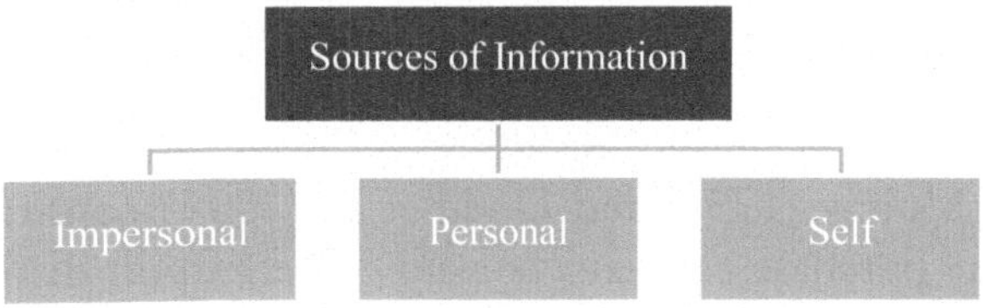

Fig. 5.4 Directional conceptual framework—information sources (*Source* Adapted from Dodd et al. [2005])

obvious when people intend to buy wine due to the complex information available (Lockshin et al., 2006). Dodd et al. (2005) concluded that wine consumers rely on three different sources: impersonal, personal and internal (i.e. themselves) (see Fig. 5.4).

5.3 Methodology

Conceptual Framework

The wine industry is extremely complicated due to wines' many influential components and customers' exposure to a wide variety of product attributes (Lockshin et al., 2006). Researchers such as Lockshin and Hall (2003) have hypothesised that wine characteristics can also be connected to the contexts in which consumers plan to consume wine. Boncinelli et al. (2019) similarly found that customers value specific wine attributes in different wine consumption settings (e.g. for personal use vs gifts). The present study's first hypothesis was based on the above findings:

H1: Specific wine characteristics are associated with each type of wine-drinking occasion.

Wine-buying decisions are significantly influenced by perceived risks (Lockshin & Hall, 2003). Bruwer and Cohen (2017) call for further investigations to examine the specific risks associated with each occasion. The current research thus included a second hypothesis written as follows:

H2: Specific risk perceptions are linked to distinct wine-drinking occasions.

In addition, Murray (1991) and Dodd et al. (2005) confirmed that, in different wine purchase contexts, consumers rely upon distinct fonts of information. This finding was incorporated into the present study's third hypothesis:

H3: Specific information sources are linked to each type of consumption occasion.

Events such as eating meals at home, dining out in a restaurant and giving gifts influence customers' attitudes and purchase behaviours in contrasting ways (DeVere et al., 1983; Goodwin et al., 1990; Oude & Van, 1995; Quester & Smart, 1998). Relationships with other consumers also affect individuals' wine choices (Hall et al., 2001). Aqueveque (2023) reports that preferences for wine labels vary for three different wine situations: barbeques with friends, gatherings with friends and dinners with a partner. These previous studies' findings were condensed into the current research's final hypothesis:

H4: Where and who can be effective segmentation variables in wine markets.

Questionnaire Design and Data Collection

This study's target population was Portuguese millennials. A survey was conducted that started by confirming if the respondents had two required characteristics. The first was these individuals' voluntary agreement to fill in the questionnaire. The second was whether they belonged to the correct age group, namely over 18 and less than 35 years old. In Portugal, 18 is the minimum age for drinking alcoholic beverages (Lancaster & Stillman, 2003). A convenience sampling procedure was followed to recruit participants, which resulted in a sample of 477 respondents.

The questionnaire included eight sections: (1) informed consent, (2) general knowledge, (3) consumption habits, (4) attributes, (5) perceived risks, (6) information sources, (7) sociodemographic information and (8) submission. The data were analysed using Pearson's chi-squared test, Cramér's V and the decision tree method of chi-squared automatic interaction detection.

5.4 Results and Discussion

Respondents' Descriptive Statistics

The sample included 147 male (31%) and 330 female (69%) respondents. Most participants were in a relationship (48%), and the majority of the remaining respondents were single (39%). The most common education levels were a bachelor's degree for 208 participants (44%) and master's degree for 165 respondents (35%). The participants' sociodemographic profile is given in Table 5.1.

The respondents most often mentioned the attributes of type of wine (73%), price (58%), brand (45%) and region (41%), while label (9%) and design (7%) were the least important. The Portuguese Generation Yers surveyed most frequently identified three risks: functional (65%), financial (55%) and social (35%). Physical (16%), time (11%) and psychological (8%) risks were selected by the fewest participants. The information sources that were most often commonly used included previous tasting (64%), relatives' advice (51%), product information (51%) and staff and/or producer's advice (37%). The respondents selected advertising (10%) and online (10%) sources far less often. Table 5.2 presents the cross-tabulation of wine attributes, perceived risks and information sources by occasion (i.e. where and who).

Hypothesis Testing

H1 postulated that specific wine characteristics are associated with each type of wine-drinking occasion. The wine attributes included were

Table 5.1 Respondents' sociodemographic profile

		Number	Percentage
Gender	Female	330	69.18
	Male	147	30.82
Age	Early millennials [18:27]	288	52.81
	Late millennials [28:35]	189	47.19
Relationship status	Single	184	38.57
	In a relationship	229	48.01
	Married	59	12.37
	Divorced	5	1.05
Current residence	Lisbon	188	39.41
	Oporto	42	8.81
	Braga	30	6.29
	Immigrant	30	6.29
	Setubal	30	6.29
	Other	157	32.91
Education	Less than secondary school	10	2.10
	Secondary school or equivalent	79	16.56
	Specialisation course	5	1.05
	Bachelor's degree	208	43.61
	Postgraduate or Master's	165	34.59
	Doctoral degree or postdoctoral	10	2.10
Current occupation	Student	139	29.14
	Working student	60	12.58
	Employee	199	41.72
	Self-employed	42	8.81
	Unemployed	29	6.08
	Other	8	1.68
Total		**477**	**100**

brand, price, origin, type of wine, design and label. The variable of who comprised five options: friends, family, partner, colleagues and myself. Three choices were given for the variable of where: restaurant, gift and home.

The chi-squared test confirmed that a significant statistical difference (i.e. a significance level of 10%) exists between all the wine attributes and

Table 5.2 Wine attributes, perceived risks and information sources by occasion

Occasion			Where			Who				
			Home	Gift	Restaurant	Myself	Friends	Family	Partner	Colleagues
Wine attributes	Price	N	169	128	191	207	157	153	151	173
		%	59	45	67	72	55	54	53	56
	Brand	N	133	155	105	99	144	134	129	129
		%	47	54	37	35	50	47	45	45
	Type of wine	N	224	191	204	211	211	213	205	197
		%	79	67	72	74	74	75	75	72
	Origin	N	122	121	111	119	116	131	118	106
		%	43	42	39	42	41	46	42	37
	Labelling	N	29	36	12	21	28	23	27	27
		%	10	13	4	7	10	8	9	8
	Design	N	15	40	8	9	26	15	23	22
		%	5	14	3	3	9	5	5	8
Perceived risks	Functional	N	188	188	181	191	182	188	188	188
		%	66	66	64	67	64	66	66	66
	Social	N	97	109	92	53	119	101	101	94
		%	34	38	32	18	42	35	35	33
	Financial	N	142	147	181	180	156	155	155	150
		%	50	51	63	63	55	55	55	53
	Physical	N	48	34	52	52	45	45	45	46
		%	17	12	18	18	16	16	16	16
	Psychological	N	25	21	25	23	22	26	26	26
		%	9	7	9	8	8	9	9	9
	Time	N	45	41	12	36	31	33	33	31
		%	16	14	4	12	11	11	11	11

(continued)

Table 5.2 (continued)

Occasion			Where			Who				
			Home	Gift	Restaurant	Myself	Friends	Family	Partner	Colleagues
Information sources	Advertising	N	32	31	21	26	27	28	26	31
		%	11	11	7	9	9	10	9	10
	Product information	N	160	158	122	145	144	150	150	149
		%	56	56	43	51	51	53	53	52
	Staff/producer's advice	N	83	86	146	116	109	106	106	106
		%	29	30	51	41	38	37	37	37
	Relatives' advice	N	163	149	128	152	150	148	148	145
		%	57	52	45	53	53	52	52	51
	Previous tasting	N	193	184	170	186	185	183	183	185
		%	68	65	60	65	65	64	64	65
	Online	N	32	36	16	24	26	28	28	29
		%	11	13	6	8	9	10	10	10

Note N = number

the who variable and between five wine attributes and the where variable. The results thus almost completely supported H1.

H2 suggested that perceived risks are linked to distinct wine consumption occasions. The measurement model thus included the functional attributes of financial, social, psychological, physical and time. The variable who again comprised five options (i.e. friends, family, partner, colleagues and myself). For the variable where, the same three choices were given (i.e. restaurant, gift and home).

The chi-squared test and Cramér's V verified that a statistically significant association exists between the who variable and social and financial risks and between the where variable and social, financial, physical and time risks. Therefore, the results provide partial support (i.e. half the associations) for H2.

H3 stated that specific information sources are linked to each type of consumption occasion. Six sources were listed: product information, advertisement, producers' information, relatives' advice, online search and previous tasting. The variable who included five options, that is, friends, family, partner, colleagues and myself. The variable where was divided into three choices: restaurant, gift and home.

The chi-squared test and Cramér's V revealed that no significant statistical connection exists between any information sources and the variable where, but clear associations appear between the information sources and the who variable. The results thus only support H3 in terms of the wine consumption occasions defined by who is involved.

H4 postulated that where and who can be considered effective segmentation variables in wine markets. The questionnaire did not include any items specifically designed to test this hypothesis, but logical deductions could be made based on the previous hypotheses' outcomes. An analysis of the 18 variables (i.e. risks, attributes and information sources) confirmed that 16 are statistically significant for the occasions defined by the where variable and 7 for the situations defined by the who variable. H4 was, therefore, considered partially verified (see Table 5.3).

Table 5.3 Overall statistical outcomes for occasions

Variables	Values	Occasion					
		Where			Who		
		χ^2 test	Cramér's V	Accept H	χ^2 test	Cramér's V	Accept H
Attributes	Price	$\chi^2 = 127.748$***	0.179	Yes	$\chi^2 = 70.812$***	0.133	Yes
	Brand	$\chi^2 = 79.230$***	0.141	Yes	$\chi^2 = 36.557$***	0.096	Yes
	Type of wine	$\chi^2 = 44.581$***	0.106	Yes	$\chi^2 = 8.625$*	0.046	No
	Region of origin	$\chi^2 = 4.974$*	0.035	No	$\chi^2 = 14.179$***	0.060	Yes
	Label	$\chi^2 = 63.542$***	0.126	Yes	$\chi^2 = 3.726$	0.031	No
	Design	$\chi^2 = 131.071$***	0.181	Yes	$\chi^2 = 23.633$***	0.077	Yes
Perceived risks	Functional	$\chi^2 = 2.003$	0.022	No	$\chi^2 = 3.060$	0.028	No
	Social	$\chi^2 = 10.285$***	0.051	Yes	$\chi^2 = 93.740$***	0.153	Yes
	Financial	$\chi^2 = 62.809$***	0.125	Yes	$\chi^2 = 18.414$***	0.068	Yes
	Physical	$\chi^2 = 19.384$***	0.070	Yes	$\chi^2 = 3.130$	0.028	No
	Psychological	$\chi^2 = 2.041$	0.023	No	$\chi^2 = 2.167$	0.023	No
	Time	$\chi^2 = 111.634$***	0.167	Yes	$\chi^2 = 1.163$	0.017	No

Variables	Values	Occasion					
		Where			Who		
		χ^2 test	Cramér's V	Accept H	χ^2 test	Cramér's V	Accept H
Information sources	Advertising	$\chi^2 = 14.271^{***}$	0.060	Yes	$\chi^2 = 2.379$	0.024	No
	Product involvement	$\chi^2 = 61.184^{***}$	0.124	Yes	$\chi^2 = 2.517$	0.025	No
	Staff/producers' advice	$\chi^2 = 184.039^{***}$	0.215	Yes	$\chi^2 = 4.782$	0.035	No
	Relatives' advice	$\chi^2 = 45.497^{***}$	0.107	Yes	$\chi^2 = 3.095$	0.028	No
	Previous tasting	$\chi^2 = 20.341^{***}$	0.071	Yes	$\chi^2 = 2.910$	0.027	No
	Online	$\chi^2 = 42.914^{***}$	0.104	Yes	$\chi^2 = 1.755$	0.021	No

Note χ^2 = chi-squared; H = hypothesis; ***, ** and * = statistically significant at the 1%, 5% and 10% level

5.5 Conclusion

The findings reveal that Portuguese Generation Yers consider price and wine type important, so they are mainly concerned about functional and financial risks. These preoccupations explain why this generation in Portugal considers having tasted wines previously to be the most reliable source of information. In addition, the empirical research conducted verified that different occasions are linked with specific wine characteristics. (e.g. Elliot & Barth, 2012), as defined by two dimensions: where and who. Social and financial perceived risks (Bruwer et al., 2013) are linked with occasions' where and who features.

The results further indicate that Portuguese millennials place different degrees of importance on wine attributes and perceived risks according to the occasion, namely when these individuals buy wine to be consumed at home with friends. This insight supports the conclusion that drinking wine in this setting with friends is these Generation Yers' most treasured occasion. However, different information sources (Nowak & Newton, 2008) become significant only in association with the where dimensions of wine occasions.

The current study also found that the variables of previous tasting, relatives' advice and product information are quite important, which suggests that market players should invest in wine tastings and wine-related events. Occasions can be paired with tactics that encourage consumption, such as unique winery experiences and bolder and more interesting information on labels. These strategies can reduce the risks associated with wine consumption, increase consumers' access to information and differentiate wines from their competitors.

The results stress that wine marketers and managers need to adopt an occasion-based segmentation approach. For instance, consuming wine with friends at home is Portuguese millennials' preferred occasion, and the majority purchase wine at supermarkets. Thus, wine businesses should seriously consider special offers (e.g. distinctive wine glasses, decanters, corkscrews and other wine products for group events), sales (i.e. to minimise financial risks) and wine tastings with qualified personnel (i.e. to lessen social risks).

The most notable finding is that where is a more effective market segmentation variable than who. The results reveal that wine consumers exhibit different behaviours depending on the situation involved. Besides confirming that bottles of wine are purchased for specific occasions (Aqueveque, 2023; Boncinelli et al., 2019; Carsana & Jolibert, 2017; Hall et al., 2001), the present research adds to the literature by considering a wider range of consumption situations including varied where and who dimensions.

Regardless of the above contributions' value, this study had some limitations that need to be considered. The convenience sampling method used may restrict the findings' generalisability, and the sample includes a higher percentage of females and individuals with a bachelor's degree, which may skew the results. Applications of the research's findings to other contexts should also be done with care because this study focused on wine consumption and buyer behaviour in Portugal.

Further studies are needed to examine more deeply some of the needs identified, such as the respondents' disregard for online information. The sample was made up of Generation Yers, so the infrequent use of online sources for information or purchasing tips was surprising.

References

Aqueveque, C. (2023). Consumers' preferences for low-priced wines' packaging alternatives: The influence of consumption occasion, gender, and age. *British Food Journal, 125*(3), 781–793.

Barber, N., Dodd, T., & Ghiselli, R. (2008). Capturing the younger wine consumer. *Journal of Wine Research, 19*(2), 123–141.

Belk, R. W. (1974). An exploratory assessment of occasional effects in buyer behaviour. *Journal of Marketing Research, 16*, 156–163.

Beninger, S., Parent, M., Pitt, L., & Chan, A. (2014). A content analysis of influential wine blogs. *International Journal of Wine Business Research, 26*(3), 168–187.

Berenguer, G., Gil, I., & Ruiz, M. (2009). Do upscale restaurant owners use wine lists as a differentiation strategy? *International Journal of Hospitality Management, 28*(1), 86–95.

Berni, P., Begalli, D., & Capitello, R. (2005). An occasion-based segmentation approach to the wine market in Denmark. *Journal of International Food & Agribusiness Marketing, 17*(1), 117–145.

Boncinelli, F., Dominici, A., Gerini, F., & Marone, E. (2019). Consumers' wine preferences according to purchase occasion: Personal consumption and gift-giving. *Food Quality and Preference, 71*, 270–278.

Brochado, A., & Martins, F. V. (2008). Aspectos metodológicos da segmentação de mercado: Base de segmentação e métodos e classificação. *RBGN: Revista Brasileira de Gestão de Negócios, 10*(27), 132–149.

Brochado, A., & Oliveira, F. (2018). Brand equity in the Portuguese vinho verde 'green wine' market. *International Journal of Wine Business Research, 30*(1), 2–18.

Brochado, A., Sena, M., & Oliveira-Brochado, F. (2022). Wine tasting and appreciation experiences at wineries. In S. Kumar Dixit (Ed.), *Routledge handbook of wine tourism* (pp. 498–506). Routledge.

Brochado, A., Stoleiu, O., & Lupu, C. (2021). Wine tourism: A multisensory experience. *Current Issues in Tourism, 24*(5), 597–616.

Bruwer, J., & Cohen, J. (2017). Restaurants and the single-serve wine by-the-glass conundrum: Risks perception and reduction effects. *International Journal of Hospitality Management, 62*, 43–52.

Bruwer, J., Fong, M., & Saliba, A. (2013). Perceived risks, risks-reduction strategies (RRS) and consumption occasions: Roles in the wine consumer's purchase decision. *Asia Pacific Journal of Marketing and Logistics, 25*(3), 369–390.

Bruwer, J., Lacey, S., & Li, E. (2009). The role of perceived risks in wine purchase decisions in restaurants. *International Journal of Wine Business Research, 21*(2), 99–117.

Carsana, L., & Jolibert, A. (2017). Is brand schematicity influencing the evaluation of product cues?: Self versus gift giving situations. *Journal of Consumer Marketing, 26*(3), 80–90.

Corduas, M., Cinquanta, L., & Levoli, C. (2013). The importance of wine attributes for purchase decisions: A study of Italian consumer's perception. *Food Quality and Preference, 28*, 407–418.

Day, S. G. (1970). *Buyer attitudes and brand choice behaviour*. The Free Press.

DeVere, P. S., Scott, D. C., & Shulby, L. W. (1983). Consumer perceptions of gift-giving occasions: Attribute saliency and structure. *Advances in Consumer Research, 10*, 185–190.

Dodd, T. H., Laverie, D. A., & Wilcox, J. F. (2005). Differential effects of experience, subjective knowledge, and objective knowledge on sources of

information used in consumer wine purchasing. *Journal of Hospitality & Tourism Research, 29*(1), 3–19.
Dowling, G., & Staelin, R. (1994). A model of perceived risks and intended risks-handling activity. *Journal of Consumer Research, 21*(1), 119–134.
Duarte, F., Madeira, J., & Barreira, M. (2010). Wine purchase and consumption in Portugal—An exploratory analysis of young adults' motives/attitudes and purchase attributes. *Ciência e Técnica Vitivinícola, 25*(2), 63–73.
Elliot, S., & Barth, J. E. (2012). Wine label design and personality preferences of millennials. *Journal of Product & Brand Management, 21*(1), 183–191.
Espejel, J., Fando, C., & Flavián, C. (2011). Antecedents of consumer commitment to a PDO wine: An empirical analysis of Spanish consumers. *Journal of Wine Research, 22*(3), 205–225.
Everard, A., & Galletta, D. F. (2006). How presentation flaws affect perceived site quality, trust, and intention to purchase from an online store. *Journal of Management Information Systems, 22*, 55–95.
Fernández, F. R., Vriesekoop, F., & Urbano, B. (2017). Social media as a means to access millennial wine consumers. *International Journal of Wine Business Research, 29*(3), 269–284.
Ferreira, C., Lourenço-Gomes, L., Pinto, L. M. C., & Silva, A. P. (2019). Is there a gender effect on wine choice in Portugal?—A qualitative approach. *International Journal of Wine Business Research, 31*(4), 618–639.
Figueiredo, I. L., Afonso, O., Ramos, P., Santos, S., & Hogg, T. (2003). *Estudo do consumidor português dos vinhos* [Study of Portuguese wine consumers]. CVRVV, IVP, AESBUC, ESB/UCP. https://www.researchgate.net/publication/311495920_Estudo_do_consumidor_Portugues_de_Vinhos. Accessed 15 April 2023.
Fountain, J., & Lamb, C. (2011). Generation Y as young wine consumers in New Zealand: How do they differ from Generation X? *International Journal of Wine Business Research, 23*, 107–124.
Goodman, S. (2009). An international comparison of retail consumer wine choice. *International Journal of Wine Business Research, 21*(1), 41–49.
Goodwin, C., Smith, K. L., & Spiggle, S. (1990). Gift giving: Consumer motivation and the gift purchase process. *Advances in Consumer Research, 17*, 690–698.
Hall, J., Lockshin, L., & O'Mahony, G. (2001). Exploring the links between wine choice and dining occasions: Factors of influence. *International Journal of Wine Marketing, 13*(1), 18–36.

Hall, J., Shaw, M., & Doole, I. (1997). Cross-cultural analysis of wine consumption motivations. *International Journal of Wine Marketing, 9*(2), 83–92.

Higgins, L., Wolf, M. M., & Wolf, J. M. (2014). Technological change in the wine market?: The role of QR codes and wine apps in consumer wine purchases. *Wine Economics and Policy, 3*(1), 19–27.

Jacoby, J., & Kaplan, L. (1972). The components of perceived risks. In M. Venkatesan (Ed.), *SV—Proceedings of the 3rd annual conference of the Association for Consumer Research* (pp. 382–393). Association for Consumer Research.

Lancaster, L., & Stillman, D. (2003). *When generations collide: Who they are, why they clash, how to solve the generational puzzle at work*. Harper Collins Publishing.

Locander, W., & Hermann, P. (1979). The effect of self-confidence and anxiety on information seeking in consumer risks reduction. *Journal of Marketing Research, 16*(2), 268–274.

Lockshin, L., & Corsi, A. M. (2012). Consumer behaviour for wine 2.0: A review since 2003 and future directions. *Wine Economics and Policy, 1*, 2–23.

Lockshin, L., & Hall, J. (2003). *Consumer purchasing behaviour for wine: What we know and where we are going*. Libraries Australia.

Lockshin, L., Jarvis, W., d'Hauteville, F., & Perrouty, J.-P. (2006). Using simulations from discrete choice experiments to measure consumer sensitivity to a brand, region, price, and awards in wine choice. *Food Quality and Preference, 17*(3–4), 166–178.

Marques, C., & Guia, A. (2015). The role of gender and occasion on the frequency of wine consumption. *Tourism & Management Studies, 11*(2), 227–234.

Mueller, S., Remaud, H., & Chabin, Y. (2011). How strong and generalisable is the Generation Y effect?: A cross-cultural study for wine. *International Journal of Wine Business Research, 23*(2), 125–144.

Murray, K. B. (1991). A test of services marketing theory: Consumer information acquisition activities. *Journal of Marketing, 55*(1), 10–25.

Nicholson, P. (1990). Gender, power and wine selection: A pilot study. *Journal of Wine Research, 11*(3), 235–243.

Nowak, L. I., & Newton, S. (2008). Using winery web sites to launch relationships with millennials. *International Journal of Wine Business Research, 20*(1), 53–67.

Orth, U. R., & Malkewitz, K. (2008). Holistic package design and consumer brand impressions. *Journal of Marketing, 72*(3), 64–81.

Oude, O. P., & Van, T. H. (1995). Perceived quality: A market driven and consumer oriented approach. *Food Quality and Preference, 6*(3), 177–183.

Quester, P., & Smart, J. (1998). The influence of consumption occasion and product involvement over consumers' use of product attribute. *Journal of Consumer Marketing, 15*(3), 220–238.

Ritchie, C., & Valentin, D. (2011). *A comparison of wine drinking behaviours in young adults in the UK and France*. Academy of Wine Business Research. https://www.academia.edu/1197026/A_Comparison_of_Wine_Drinking_Behaviours_in_Young_Adults_in_the_UK_and_France. Accessed 15 April 2023.

Sanches, C. (2013). *Comportamentos de consumo de vinho: Envolvimento, estilos de vida, risco e atributos* [Wine consumers' behaviours: Involvement, lifestyles, risks and attributes] (Unpublished master's dissertation). Instituto Superior Politécnico de Viseu, ESE.

Sandell, R. G. (1968). The effects of attitudinal and occasional factors on reported choice behaviour. *Journal of Marketing Research, 4*, 405–408.

Schiffman, L., O'Cass, A., Paladino, A., D'Alessandro, S., & Bednall, D. (2011). *Consumer behaviour* (5th ed.). Pearson Australia.

Sellers, R., & Nicolau, J. L. (2023). Satisfaction and expenditure in wineries: A prospect theory approach. *Journal of Hospitality & Tourism Research, 47*(2), 354–374.

Sogari, G., Corbo, C., Maccobi, M., Menozzi, D., & Mora, C. (2015). Consumer attitude towards sustainable-labelled wine: An exploratory approach. *International Journal of Wine Business Research, 27*(4), 312–328.

Thach, L. (2009). Wine 2.0—The next phase of wine marketing?: Exploring US winery adoption of Wine 2.0 components. *Journal of Wine Research, 20*(2), 143–157.

Williamson, P. O., Lockshin, L., Francis, I. L., & Loose, S. M. (2016). Influencing consumer choice: Short and medium term effect of country of origin information on wine choice. *Food Quality and Preference, 51*, 89–99.

6

Exploring Emerging Chinese Wine Tourists: Does Gender Matter?

Stella Kladou, Ahmet Usakli, and Kyuho Lee

6.1 Introduction

The black box of consumer behavior has been the cornerstone of numerous studies in tourism and wine research alike. Researchers have developed typologies to help practitioners better respond to tourists' and wine consumers' needs and characteristics. Toward this direction, scholars have suggested segmentation variables and identified the role of attributes unique to wine (e.g. involvement) (e.g. Alebaki & Iakovidou, 2011; Alonso et al., 2022), with a few studies providing some insights into wine consumers and tourists from emerging wine destinations and New World wine countries (e.g. Banks & Overton, 2010; Lee et al.,

S. Kladou (✉)
Hellenic Mediterranean University, Heraklion, Greece
e-mail: stellakladou@hmu.gr

A. Usakli
Boğaziçi University, Istanbul, Turkey

K. Lee
Sonoma State University, Rohnert Park, CA, USA

J. Martínez-Falcó et al. (eds.), *Strategic Management in the Wine Tourism Industry*,
https://doi.org/10.1007/978-3-031-54837-6_6

2022; Zhao, 2015). Taken the particularities of Asian cultures and societies, and the recently recognized gender role on wine consumption (Lee et al., 2019), this study explores Chinese wine tourists' perceptions of winery service quality, winery satisfaction and winery loyalty. In specific, the study is designed to answer the following research questions (RQs):

- RQ1: Do Chinese wine tourists' perceptions of winery service quality differ based on gender?
- RQ2: Do Chinese wine tourists' perceptions of winery satisfaction and winery loyalty differ based on gender?

Analyzing the responses of 258 Chinese wine tourists visiting Yantai, China, to a structured questionnaire, reveals that Chinese male wine tourists' winery satisfaction and winery loyalty are significantly higher than Chinese female wine tourists. Moreover, in terms of winery service quality, Chinese male wine tourists have significantly more positive perceptions of winery staff behavior. Building on our findings, this chapter discusses gender roles in relation to the Chinese wine tourist. As a result, scholars gain an insight on segmentation variables applicable to Chinese wine tourists. Similarly, winery managers become more informed on what to prioritize for their product and service to more effectively address this growing wine tourism market.

6.2 Theoretical Foundation

Winery service quality is a well-established construct in the literature, commonly embracing the physical attributes of a winery, such as winery architecture, facilities, vineyards and natural landscape (Alonso, 2014; Gunasekar et al., 2022; Lee & Ha, 2014; Quintal et al., 2015), winery staff behavior and hospitality (e.g. Lee et al., 2021) and, of course, the quality of wine and wine tastings (e.g. Byrd et al., 2016; Qiu et al., 2013). Similarly, Byrd et al. (2016) have already discussed how physical attributes in specific are important in order to boost satisfaction. Other researchers link the servicescape with satisfaction and behavioral intentions, particularly loyalty and the intention to return and/or recommend the specific winery and/or wine destination (e.g. Lee et al., 2022; Shin & Nicolau, 2022). In response, consumer insights relevant to winery service quality, satisfaction and loyalty from both a product/corporate and a leisure/ tourism aspect are important.

Visiting different wine regions allows wine consumers to taste diverse wines, experience the terroir of each wine region and learn the history and tradition of the respective wine region. Along with the increased wine consumption in China (Lee et al., 2022), wine tourism has emerged as a leisure activity among Asian and Chinese wine consumers as well (Qiu et al., 2013), who, depending on the circumstances (e.g. the pandemic), may prefer nearby or internationally established wine destinations, see wine as a lifestyle product and are increasingly more interested in visiting wine regions and wine museums.

In response, extant literature has already explored some similarities and differences with Asian, and specifically Chinese, wine consumers and tourists, in relation to the quality, satisfaction and loyalty, thereby providing some input to extant scales. For instance, cultural aspects (e.g. the interpretation of eye contact with winery staff for Western versus Asian wine tourists) may influence quality evaluations per se (Lee et al., 2013), as well as the motives behind wine tourism (Lee et al., 2022; Qiu et al., 2013). Chinese wine consumers believe that wine is a product they need to learn to truly appreciate the quality of the product, while Western wine consumers commonly view wine as part of their culture and everyday life (Salustri et al., 2022). Thus, prior research (Lee et al., 2016) reported that the Chinese wine tourists value supplementary service provided at the cellar door (i.e. friendliness of the winery staff, atmosphere and landscape of the winery) more, as learning wine is one of the major goals for the Chinese wine tourists to involve with wine tourism (Lee et al., 2016). Researchers assert that experiencing the culture and tradition of the wine region through wine tourism is key to wine tourists as wine is ingrained into local cuisine, culture and tradition (Zhao, 2015).

Over the years, wine emerged as a popular alcoholic drink among Asian consumers. Studies have reported that this growing number of Asian consumers has contributed to the increased sales of global wine companies, particularly the wine companies located in old wine countries such as France (Zhao, 2015). The accelerating globalizing economy, rapidly increasing middle-class consumers and the urbanization in many Asian countries have contributed to the phenomenon of increased wine consumption in Asia (Lee et al., 2022). Among all the Asian countries, China has been one of the most important wine markets in Asia; wine consumption among Chinese consumption is particularly notable (Lee et al., 2016). Researchers postulate that Chinese consumers view wine as a symbol of high-class Western lifestyle and social status (Maguire & Zhang, 2017): For Chinese consumers, drinking a glass of wine from Bordeaux, France reveals not only somebody's economic status but also a particular, well-appreciated social and cultural capital. Furthermore, wine is a very popular gift among Chinese consumers, a fact that has elevated the sales of Western wines in China, taken the long tradition of exchanging gifts in China during the holiday seasons (Maguire & Zhang, 2017).

Interestingly, female wine drinkers are one of the major drivers for the sales of wine in China and other Asian countries (Lee et al., 2019). This is an interesting phenomenon as Asian female consumers have been restricted from alcoholic drinks as this deviates from the traditional female gender role. While China is a masculine country (Hofstede, 2023) and drinking alcohol is considered a primarily male activity, female Chinese consumers also drink wine and visit wineries (Zhao, 2015). In fact, researchers (Lee et al., 2019) report that Asian female consumers view wine as a way to empower one's self-identity. In spite of the increasing wine consumption among Chinese female consumers, little research has sought to understand Chinese female wine consumers. In response, this study focuses on Chinese wine tourists and investigates the role of gender for winery service quality, satisfaction and loyalty.

6.3 Method

Since a quantitative research design best fits the research objective, this study employs the Nordic School perspective to assess winery service quality and asks respondents to evaluate the winery experience based on physical quality (six items), staff behavior (nine items) and wine tasting (five items) (Ekinci, 2001; Ekinci et al., 2008). To measure winery satisfaction, seven items were used to capture both emotional satisfaction and evaluative satisfaction (Carlsen, 2011; Charters et al., 2009), whereas winery loyalty was measured using four behavioral intentions (see Appendix for the full list of measures). We used a seven-point Likert scale to measure all constructs. The research survey was originally in English, but, in order to facilitate respondents, was translated and back-translated by bilingual Chinese wine professors. The pilot study confirmed that the research tool was appropriate, accurate and ready to use.

Data collection took place at Changyu winery, in Yantai, China. Yantai is one of the most important wine regions in China and Changyu one of the oldest wineries in the area, and the largest wine producer in China (Qiu et al., 2013). Winery visitors were invited to complete the self-administered questionnaire once their winery visit was complete. More than 300 visitors responded. Yet, eliminating questionnaires with excessive missing data, outliers and random responses, fed our analysis with a total of 258 questionnaires. For the analysis, we used SPSS to conduct independent samples t-test and respond to our research questions.

6.4 Findings

Demographic Profile of Respondents

In this study, a convenience sample of 258 Chinese wine tourists visiting Yantai, China, was analyzed. Out of the 258 Chinese wine tourists surveyed, 53.9% were women. Additionally, 60.9% fell into the age range of 30–50 years old, and 77.5% held university degrees. The majority of respondents (85.7%) identified themselves as leisure travelers. A significant proportion of respondents (60.5%) had previously visited Yantai, with a majority of them (71.7%) staying for a duration of five days or more. Almost half of the respondents (48.1%) had been consuming wine for a minimum of ten years, and on average, they visited five wineries per year. Hence, the respondents possessed substantial wine tourism experiences.

Reliability Assessment and Research Question Analysis

Prior to testing the research questions, we first assessed the reliability of the study variables (i.e. winery satisfaction, winery loyalty and the three dimensions of winery service quality). All variables showed strong internal consistency (i.e. Cronbach's alpha coefficients > 0.85). Having established the reliability of the study variables, we then computed the mean scores of each variable to test the research questions.

RQ1 investigated if any statistically significant differences existed between the perceived winery service quality of Chinese female and male wine tourists. To do this, an independent samples t-test was performed on the mean scores of three dimensions of winery service quality (see Table 6.1). Of the three dimensions of winery service quality (i.e. physical quality, staff behavior and wine wasting), only perceptions of *staff behavior* were found to significantly differ based on gender. Specifically, Chinese male wine tourists ($\bar{x} = 5.48$) scored significantly higher than Chinese female wine tourists ($\bar{x} = 5.15$) on staff behavior ($p = 0.01$). This reveals that Chinese male wine tourists have significantly more positive perceptions than Chinese female wine tourists with regard to winery staff behavior. Additionally, we calculated the Cohen's d index to determine the effect size. The magnitude of this statistically significant difference between Chinese female and male wine tourists ($d = 0.32$) was found to be small (Cohen, 1977). No statistically significant differences were found between Chinese female and male wine tourists for *physical quality* and *wine tasting* dimensions.

RQ2 examined whether significant differences existed between Chinese female and male wine tourists in terms of perceived winery satisfaction and winery loyalty. To test RQ2, we carried out an independent samples t-test on the mean scores of winery satisfaction and winery loyalty. The results are presented in Table 6.2. The results indicate

Table 6.1 Comparison of winery service quality dimensions between Chinese female and male wine tourists

Dimensions of winery service quality	Mean score		*t*-value	Cohen's *d*
	Female (n = 139)	Male (n = 119)		
Physical quality	5.40	5.49	0.672	0.32
Staff behavior	5.15	5.48	2.606**	–
Wine tasting	5.18	5.36	1.331	–

**$p < 0.01$

Table 6.2 Comparison of winery satisfaction and winery loyalty between Chinese female and male wine tourists

	Mean score		*t*-value	Cohen's *d*
	Female (n = 139)	Male (n = 119)		
Winery satisfaction	5.26	5.53	2.152*	0.27
Winery loyalty	5.14	5.50	2.640**	0.33

*$p < 0.05$, ** $p < 0.01$

that Chinese wine tourists' perceptions of both winery satisfaction and winery loyalty significantly differ based on gender. Specifically, Chinese male wine tourists' winery satisfaction ($p = 0.032$) and winery loyalty ($p = 0.009$) were found to be significantly higher than Chinese female wine tourists. Additionally, Cohen's d effect sizes demonstrated that the magnitude of differences between Chinese female and male wine tourists were small for both winery satisfaction ($d = 0.27$) and winery loyalty ($d = 0.33$).

6.5 Discussion

The importance of segmentation remains vital in the wine (tourism) literature, with recent studies emphasizing variables unique to the wine industry, such as involvement, motivation or wine lifestyle (e.g. Alonso et al., 2022). Still, demographics cannot be neglected in an effort to recognize the wine tourist that may be more important in the long term. Focusing specifically on gender, in 'New World' countries such as New Zealand and Australia, wine tourists are typically female, whereas in 'Old World' European countries, most wine tourists are men (Alebaki & Iakovidou, 2011). In other 'New World' countries, however, the Confucian philosophy, widespread especially in South-East Asia, emphasizes gender roles and specific lifestyles and, hence, restricts women from the consumption of non-traditional products, such as wine. In response to such traditional cultural aspects being blended with contemporary trends (e.g. the rise of the middle-class, the importance of economic and socio-cultural status), Lee et al. (2019) described how South Korean female consumers create a unique wine consumption culture. Taking such cultural similarities among Asian countries and Confucian cultures and the significance of segmentation variables for businesses, including wineries, this study set off to investigate the importance of this 'traditional' variable of gender for specific important constructs for winery management, namely service quality, satisfaction and loyalty, and explored the case of Chinese wine consumers instead. Unlike Lee et al.'s (2019), this study explored Chinese wine consumers' perceptions while they were performing a wine leisure activity (i.e. visiting a winery).

Of the three dimensions of winery service quality (i.e. physical quality, staff behavior and wine tasting), only perceptions of staff behavior significantly differ based on gender. Specifically, Chinese male wine tourists have significantly more positive perceptions than female wine tourists with regard to winery staff behavior, thereby suggesting winery staff could diversify their behavior and expected output based on the gender of the tourist they encounter with. No statistically significant differences were found for physical quality and wine tasting dimensions, which contradicts early findings in another New World country context (namely New Zealand), according to which female winery visitors were more likely to enjoy the ambiance of the winery (Mitchell & Hall, 2001).

With regard to the second research question, Chinese wine tourists' perceptions of both winery satisfaction and winery loyalty significantly differ based on gender. Specifically, Chinese male wine tourists' winery satisfaction and winery loyalty are significantly higher compared to Chinese female wine tourists. Greater satisfaction with the wine tourism experience can help the winery and wine region inspire return visits (Roberts & Sparks, 2006), and support demand in the long term through brand and customer loyalty (Alebaki & Iakovidou, 2011; Getz, 2000). Given the importance which wine tourism attributes to *attracting, educating and building long-term emotional and social ties with their guests* (Sigala & Robinson, 2019: 2), gender emerges as an important parameter to support such efforts in the Chinese context.

The results of the study partly align with previous study on the impacts of gender on Asian consumers' service preferences (Lee et al., 2013). According to this earlier work, Asian female consumers tend to have higher expectations on service quality, particularly on the dimension of intangibility such as friendliness, kindness and attentiveness (Lee et al., 2013). The results of this study also suggest that wineries need to develop a competitive service strategy to enhance the overall experiences of the Chinese female wine tourists. For example, wineries might provide more personalized and attentive service to Chinese female wine tourists at the winery who might have higher service expectation than that of Chinese male tourists. The passion to learn about wine during the winery visit among Chinese wine tourists is high (e.g. Qiu et al., 2013). Thus, providing more educational information such as the terroir of the wine region and the winery to female Chinese winery visitors will help increase the satisfaction of all Chinese wine tourists.

In sum, this is one of the first studies exploring the impacts of gender on Chinese wine tourists' satisfaction, loyalty and service quality perceptions when visiting a winery. Future studies might look more into the impact and role of gender among winery tourists and wine consumers using samples from Western countries and other Asian countries to validate the results of the current study and proceed to strategy recommendations valuable for wineries addressing multi-cultural target groups.

Appendix: Measurement of the Variables

Respondents were asked to answer the following questions based on their winery visit.

Winery service quality
Physical quality
The décor of the winery was visually appealing
The winery was tidy
The winery provided a comfortable facility
The winery provided a good landscape and garden
The winery vineyard is attractive physically
Winery staff appeared neat
Staff behavior
Staff was willing to help
Staff was responsive
Made me feel like a special person
Staff seemed to anticipate what I wanted
Staff had visitors' best interests at heart
Staff understands the specific needs of visitors
Knowledge of staff was appropriate
Behavior of staff instilled confidence
Felt secure in dealing with winery staff
Wine tasting
Range of wines
Convenience of tasting times
Quality of wine tasted
Acceptable waiting time
Side dish (or finger food) provided for wine
Winery satisfaction
Thinking about visiting this winery makes me happy
I feel a pleasant feeling when I think about visiting this winery
This winery met my expectations
I made the right choice in visiting this winery
Visiting this winery gave me high satisfaction
Visiting this winery has answered my wishes
I am pleased to visit this winery
Winery loyalty
I'll keep visiting this winery
I'll recommend this winery to others
This winery will be my first choice in the future
I intent to buy the wine products produced from this winery

All items were measured on a seven-point Likert-type scale

References

Alebaki, M., & Iakovidou, O. (2011). Market segmentation in wine tourism: A comparison of approaches. *Tourismos: An International Multidisciplinary Journal of Tourism, 6*(1), 123–140.

Alonso, A. D. (2014). Wine cellar experiences in the Southeastern United States: Educating the winery visitor on Muscadine wines. *Journal of Foodservice Business Research, 17*(1), 1–18.

Alonso, A. D., Bressan, A., Vu, O. T. K., Ha Do, L. T., Garibaldi, R., & Pozzi, A. (2022). How consumers relate to wine during COVID-19—A comparative, two nation study. *International Journal of Wine Business Research, 34*(4), 590–607.

Banks, G., & Overton, J. (2010). Old World, New World, Third World? Reconceptualising the worlds of wine. *Journal of Wine Research, 21*(1), 57–75.

Byrd, E. T., Canziani, B., Hsieh, Y., Debbage, K., & Sonmez, S. (2016). Wine tourism: Motivating visitors through core and supplementary services. *Tourism Management, 52*, 19–29.

Cohen, J. (1977). *Statistical power analysis for the behavioral sciences* (Revised ed.). Academic Press.

Ekinci, Y. (2001). The validation of the generic service quality dimensions: An alternative approach. *Journal of Retailing and Consumer Services, 8*(6), 311–324.

Ekinci, Y., Dawes, P. L. & Massey, G. R. (2008). An extended model of the antecedents and consequences of consumer satisfaction for hospitality services. *European Journal of Marketing, 42*(1), 35–68.

Gunasekar, S., Das, P., Dixit, S. K., Mandal, S., & Mehta, S. (2022). Wine-experienscape and tourist satisfaction: Through the lens of online reviews. *Journal of Foodservice Business Research, 25*(6), 684–701.

Getz, D. (2000). *Explore wine tourism: Management, development, destinations*. Cognizant.

Hofstede, G. (2023). *Hofstede Insights for China*. https://www.hofstede-insights.com/country-comparison/china/. Accessed May 3, 2023.

Lee, K., & Ha, I. (2014). From Biltmore Estate to Biltmore winery. *Journal of Foodservice Business Research, 17*(5), 429–438.

Lee, K., Kladou, S., Usakli, A., & Shi, Y. (2022). Inspiring winery experiences to benefit destination branding? Insights from wine tourists at Yantai, China. *Journal of Hospitality and Tourism Insights, 5*(1), 116–137.

Lee, K., Madanoglu, M., Hanson, S. W., & Ko, J.-Y. (2019). The gateway to consumption freedom through a communal glass of wine: South Korean female wine consumers and wine consumption community. *International Journal of Wine Business Research, 31*(3), 303–326.

Lee, K., Madanoglu, M., & Ko, J. (2013). Developing a competitive international service strategy: A case of international joint venture in the global service industry. *Journal of Services Marketing, 27*(3), 245–255.

Lee, K., Madanoglu, M., & Ko, J. (2016). Exploring key service quality dimensions at a winery from an emerging market's perspective. *British Food Journal, 118*(12), 2981–2996.

Maguire, J. S., & Zhang, D. (2017). Shifting the focus from consumers to cultural intermediaries: An example from the emerging Chinese fine wine market. *Consumer Culture Theory, 18*, 1–27.

Mitchell, R., & Hall, C. M. (2001). The influence of gender and region on the New Zealand winery visit. *Tourism Recreation Research, 26*(2), 63–75.

Qiu, H. Z., Yuan, J. J., Ye, B. H., & Hung, K. (2013). Wine tourism phenomena in China: An emerging market. *International Journal of Contemporary Hospitality Management, 25*(7), 1115–1134.

Quintal, V. A., Thomas, B., & Phau, I. (2015). Incorporating the winescape into the theory of planned behaviour: Examining 'new world' wineries. *Tourism Management, 46*, 596–609.

Roberts, L., & Sparks, B. (2006). Enhancing the wine tourism experience: The customers' viewpoint. In J. Carlsen & S. Charters (Eds.), *Global wine tourism: Research management and marketing*. CABI Publishing.

Salustri, A., Cocco, V., & Mawroh, H. (2022). Mixing culture, food and wine, and rural development with the Covid-19-related crisis of the tourism industry in Italy. *Journal of Foodservice Business Research, 25*(6), 702–724.

Shin, S., & Nicolau, J. L. (2022). Identifying attributes of wineries that increase visitor satisfaction and dissatisfaction: Applying an aspect extraction approach to online reviews. *Tourism Management, 91*, 104528. https://doi.org/10.1016/j.tourman.2022.104528

Sigala, M., & Robinson, R. N. S. (2019). Introduction: The evolution of wine tourism business management. In M. Sigala & R. N. S. Robinson (Eds.), *Management and marketing of wine tourism business: Theory, practice, and cases* (pp. 1–21). Palgrave Macmillan.

Zhao, J. (2015). Winery tourism in China. In K. Lee (Ed.), *Strategic winery tourism and management* (pp. 127–142). Apple Academic Press.

7

A Systematic Review of Wine Tourism and Customer Behavior Using Bibliometric Analysis

Mobina Mohammadi, Jorge Linuesa Langreo, Pablo Ruiz-Palomino, and Ricardo Martínez-Cañas

7.1 Introduction

Wine tourism is becoming increasingly relevant in Spain, evidenced by the exponential growth in visits to wineries (Treloar et al., 2004). The report published by Statista (2023) on the number of visitors to wineries and wine museums in Spain from 2019 to 2021 shows that after the

M. Mohammadi · J. L. Langreo (✉) · P. Ruiz-Palomino · R. Martínez-Cañas
Department of Business Administration, University of Castilla-La Mancha, Ciudad Real, Spain
e-mail: Jorge.Linuesa@uclm.es

M. Mohammadi
e-mail: Mobina.Mohammadi@alu.uclm.es

P. Ruiz-Palomino
e-mail: Pablo.Ruiz@uclm.es

R. Martínez-Cañas
e-mail: Ricardo.Martinez@uclm.es

J. Martínez-Falcó et al. (eds.), *Strategic Management in the Wine Tourism Industry*,
https://doi.org/10.1007/978-3-031-54837-6_7

coronavirus pandemic (COVID-19), the number of wine tourists in Spain recovered notably from the beginning of 2021 (see Fig. 7.1).

The importance of wine tourism has also been evidenced at an academic level as different types of research related to wine tourism have been addressed, such as customer behavior (e.g., Esau & Senese, 2022; Santos et al., 2021; Sigala, 2020) and segmentation of winery cellar doors (Santos et al., 2020; Thomas et al., 2013), among others. For this reason, we consider it necessary to carry out a bibliometric analysis that helps us understand the knowledge map that exists regarding wine tourism and customer behavior.

With the rapid development of the tourism industry and competition, customer behavior has become a key topic. By using bibliometric analysis to study wine tourism behavior, we can identify the level of development of the literature on wine tourism and fill the bibliometric gap on this topic. Wine tourism is currently one of the most important and promising types of tourism, since it is linked to the new consumption patterns of tourists today (Festa et al., 2020) that revolve around the importance of the experience, as well as the shorter duration of visits and their greater frequency. A hedonistic perspective on wine tourism

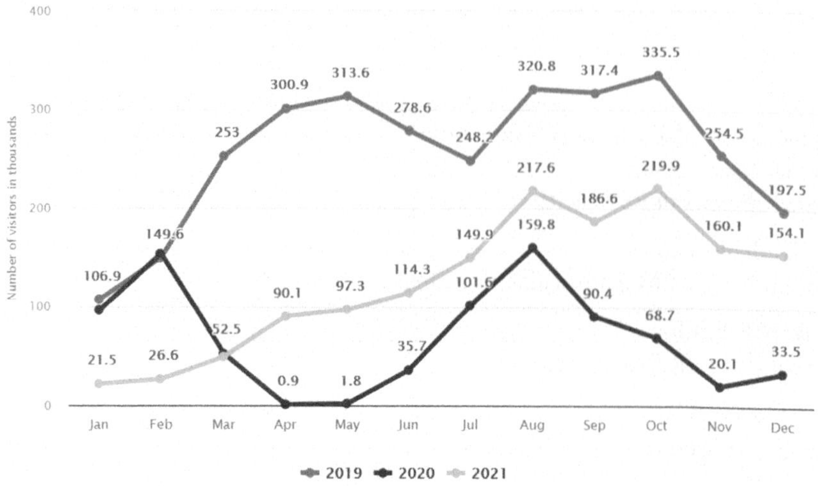

Fig. 7.1 Number of visits to Spanish wine museums and wineries (2019–2021) (*Source* Statista [2023])

consumption seems justified because the nature of wine tourism involves indulging the senses in the wine product and its immediate aesthetic surroundings (Charters et al., 2009). As a result, food and wine can emerge as the main reason to travel and visit a country, rather than a secondary activity (López-Guzmán et al., 2012).

Bibliometric analysis is increasingly being used as a method to evaluate the effectiveness of existing research and helps academics to understand the development of the literature in a certain field of study such as the one here analyzed (the relationship between customer behavior and wine tourism). The metrics used are based on bibliographic databases that are primarily aimed at information retrieval and systematization (Hood & Wilson, 2003). Therefore, since there is currently no review of this literature, this paper introduces the bibliometric analysis of wine tourism and customer behavior, thus filling an important gap in the literature. In this way, future research can use bibliometric analysis to complete the intellectual framework of wine tourism and identify developing trends in article and journal output, patterns of cooperation and research components (Donthu et al., 2021a, 2021b; Verma & Gustafsson, 2020).

7.2 Wine Tourism and Customer Behavior

Wine tourism refers to travel associated with a winery or wine region, as well as direct tourism opportunities for sales and marketing (Getz, 2000). Wine tourism includes visits by customers to vineyards, wineries, wine festivals and wine exhibitions and refers to the tasting of wine as the main driver for such visits (Hall et al., 2000). Thus, a leisure activity in which enjoyment and personal development is searched is very characteristic of wine tourism (Beames, 2003).

Indeed, visits to wineries are typically perceived as leisure activities or sources of entertainment, rather than simply as trips to purchase wine (Bruwer & Alant, 2009). Winery visitors expect an experience that goes beyond the simple purchase of a bottle of wine as they seek hedonistic and indulgent wine-related experiences (Bruwer & Alant, 2009). In this way, when the consumer has completed the whole experience, they will feel more capable and competent in appreciating the different products associated with wine (Thach & Olsen, 2003). Indeed, according to Bruwer (2002), the experience at the cellar door definitely influences the return and the spread of positive word-of-mouth (by the visitor of the winery) about the establishment. Roberts and Sparks (2006) find that first impressions of a winery, reflecting reactions to the initial contact, are significant, and if the service is outstanding this may be the deciding factor for subsequent visits.

Wine tourism is increasingly popular today among different segments of people. There are up to five segments, according to Charters and Ali-Knight (2002): wine lovers, wine connoisseurs, wine enthusiasts, wine novices and wine amateurs. Along the same lines, Hall et al. (2009) suggest three segments of wine tourists: wine lover, wine interested and curious tourist. These various classifications highlight the segmentation of the market, as the purpose of the trip for each tourist is different. Wine tourism is an essential and easy way for the wine industry to connect with customers by giving them the opportunity of direct experience with the grape (Hall et al., 2000).

7.3 Bibliometric Study of Wine Tourism and Customer Behavior

Data: Building the Database

Based on the importance of wine tourism as an industry, we run a bibliometric analysis to determine the lines of research over the last twenty years (2002–2022); studies prior to 2002 are residual. In order to carry out a bibliometric analysis, it is necessary to implement a holistic technique to analyze the structure of knowledge relating wine tourism to customer behavior. Thus, the knowledge map was created using VOSviewer, with the Web of Science Core Collection (WoS) being used as a source of information.

We combined two different set of keywords to reflect the wine tourism and customer behavior concepts. In addition, we set two restrictions, one related to *limiting the search* to *articles* only and the other related *to restricting the search to articles in English.* The search returned 171 documents using the following syntax: "wine tourist*" OR "wine tourism*" AND "innovation*" AND "customer behavior"* AND "consumer behavior*."

Figure 7.2 shows the 171 publications selected from the Web of Science Core Collection, most of which belonged to the category of Hospitality, Leisure, Sport and Tourism. Specifically, 90 papers were found in this area of knowledge. Additionally, the sample included 28 papers in the area of agronomy and 26 papers on management. In addition, several of the most frequent published articles were related to tourism and the wine sector (e.g., International Journal of Wine Business Research, Tourism Management, Journal of Travel Research, Tourism Marketing, International Journal of Contemporary Hospitality Management, International Journal of Culture Tourism and Hospitality Research). Additionally, Fig. 7.3 revealed that no articles were published before 2002, with an exponential growth in the number of publications beginning in 2008. This effect accelerated from 2016 until the COVID-19 pandemic, which had a significant effect on the number of publications in 2020 and 2021. However, the greatest percentage of articles were published in the period between 2019 and 2022. In terms of citations, the evolution is similar, with 2022 being the year with the highest number of citations. Constant growth can be observed between 2007 and 2022, especially from 2016.

Bibliographic Coupling

VOSviewer finally analyzed the 171 articles, performing an additional filter to reduce the number of groups related to one another (Fig. 7.4). The network was analyzed, yielding a structure with 6 groups. However, after inspecting these clusters, we added an additional filter in VOSviewer with the aim of minimizing the number of clusters (van Eck & Waltman, 2010).

Consequently, one of the clusters was eliminated due to its low number of articles and the lack of relationship between these and the rest of the knowledge map. Therefore, 5 clusters were finally obtained, which

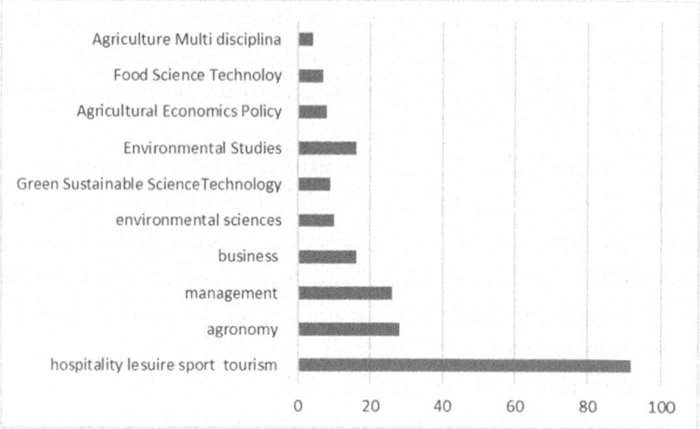

Fig. 7.2 Number of articles in each area (*Source* Web of Science)

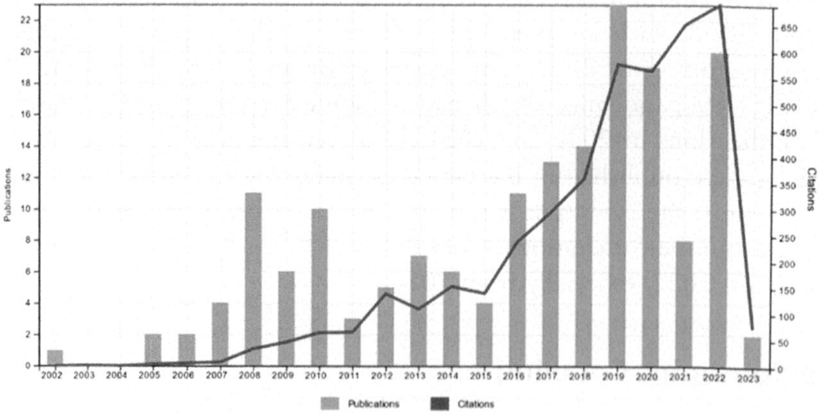

Fig. 7.3 Number of citations and publications (*Source* Web of Science)

are represented by different colors in Fig. 7.4. In this figure, one can see lines that are thicker than others such that the thickener the line is, the more depth or intense is the relationship between the nodes in question. After analyzing the articles presented in each of the clusters, we labeled each cluster and described its main characteristics and properties within the network (see Table 7.1).

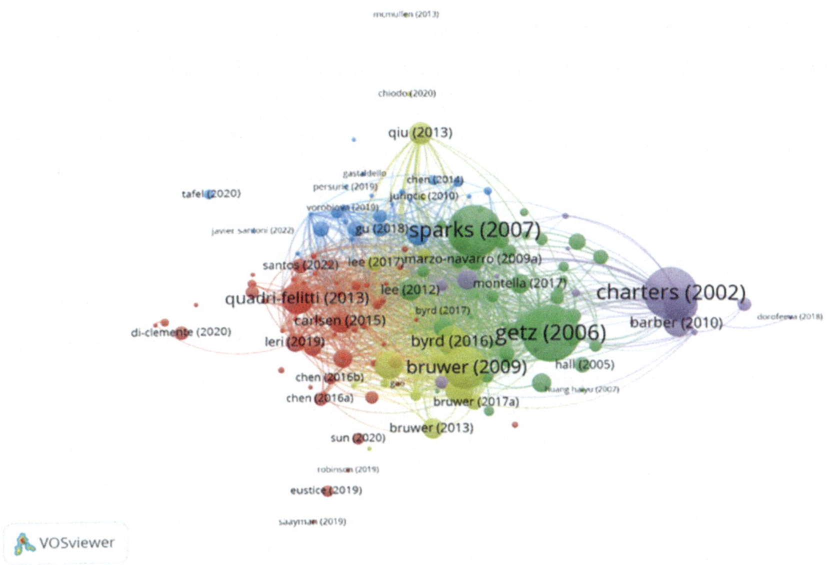

Fig. 7.4 Bibliographic coupling network

Table 7.1 Characteristics of clusters

Color	Name	Number of documents	Average publication year	Average citations (WOS)
Red	Making wine tourism experiences	53	2019	68.43
Green	Developing and analyzing wine tourism behavior	49	2011	51.70
Yellow	Wine tourist perspective and winescape	25	2016	61.24
Blue	Wine tourist segmentation and analysis of cellar doors	25	2018	49.34
Purple	Wine tourism and services	15	2013	52.78

Source Own preparation

As recommended in the literature, a bibliographic coupling analysis was carried out, as this technique has been described as critical to generate a knowledge structure of any research topic. In specific, we followed the recommendations raised by Zupic and Čater (2015) and particularly used VOSviewer for our graphical analysis of the results.

Network Analysis

As shown (Fig. 7.4), two clusters prevail in the network, as they are located in the center of it. One is red (making wine tourism experiences) and the other is green (developing and analyzing wine tourist behavior). These two clusters are transversal in nature and serve to connect the rest of the network.

The blue cluster (case study of wine tourist, segmentation and analysis of cellar doors) was far from center, while the yellow cluster (wine tourist perspective and winescape) is extensively connected to the rest of clusters. In fact, the yellow cluster is closely connected with the red and green cluster. The purple cluster (wine tourism and services) is far from the center but contains a small number of articles connected to the green and yellow clusters. It is important to note that the green cluster builds bridges with several other clusters, specifically this cluster is a bridge between the blue and yellow clusters. The yellow and green clusters are distributed across the network. The red one (making wine tourism experiences) is certainly close and very connected with the blue cluster, yet it is strongly connected with the yellow and green cluster as well. In contrast, the purple one (wine tourism and services) is far from the center but is connected to the green and yellow clusters.

All three circumstances of the green, yellow and red clusters define the centrality and density of them. Thus, the place that the blue cluster occupies in the graph (wine tourist, segmentation and analysis of cellar doors), as outside the center of the network, is an implication of presenting very low centrality, and only relatively medium to high density.

The different clusters were analyzed revealing the main themes—when there is a high level of internal growth and strong connection with other clusters—and the isolated themes—when there is a high degree of internal growth but a weak connection with the rest of the declining themes. Moreover, the latter are not yet sufficiently applicable to the field.

The **red cluster** (making wine tourism experiences) is the cluster that presents a high level of density; in specific, it presents the highest density level. Thus, we can say that this group of research articles has a focus on a very unique topic and is likely the youngest cluster of all: WOS shows the average publication year to be 2019. The large number of documents and citations in this group is one of the reasons for this group having important trends and themes. The main themes of the articles in this group focus on several aspects: (a) experience in wine tourism (e.g., Zhang & Lee, 2022); (b) the effects that the wine tourism experience has on the behavior of consumers, and especially, on their co-creation and the perceived value of such experience (e.g., Chen et al., 2016; Hernández et al., 2022; Santos et al., 2020, 2021); (c) tourists' evaluation of a comprehensive tourist experience (e.g., Fernandes & Cruz, 2016; Martins et al., 2017); and (d) the search for emotions and sensations by tourists and their relationship with culture (Björk & Kauppinen-Räisänen, 2016).

The **green cluster** (developing and analyzing the behavior of wine tourism) is the oldest (based on WOS), with average year of publication being 2011; consumer, demand and behavior are the most repeated topics. One of the studies with the greatest centrality is that by Asero and Patti (2011), which analyzes the link between products linked to wine tourism and consumer behavior (Yuan et al., 2008). Other articles included in the cluster analyze how wine tourist behavior can vary depending on factors such as age, gender, wealth and education (Galloway et al., 2008; Getz & Brown, 2006). They indicate the importance of analyzing tourist demand, as well as understanding the tastes and behavior of visitors, evidencing that wineries need to better know and apprehend wine tourists' traits and motivations (Charters & Ali-Knight, 2002).

The **yellow cluster** (wine tourist perspective and winescape) reveals that the vast majority of brand loyalty research documents have only looked at how customer behavior relates to the tasting room experience (Bruwer & Alant, 2009). Bruwer et al. (2013) discovered that the wine tourist has a mostly hedonistic experience when they move through the winescape, which consists of visually pleasant locations of both a physical and natural nature. Future researchers on this topic should delve into distinguishing and profiling wine tourists and forecasting their behavioral intentions (Thomas et al., 2018).

The **blue cluster** (wine tourism segmentation and analysis of cellar doors) includes studies that analyze wine tourism segmentation and its impact on the economy of wineries and the region. The articles in this group are far from the center and the group's average year of publication is 2018. This cluster contains isolated topics that have significant intellectual development but a weak connection to the rest of the network. The articles with the greatest centrality are those by Brunner and Siegrist (2011), Charters and Ali-Knight (2002), Szolnoki (2018) and Tafel and Szolnoki (2020). Specifically, Brunner and Siegrist (2011) suggest nine segmentation variables: quality, consumption, risk reduction, occasion, culture, behavior, participation, geography and lifestyles.

In the **purple cluster** (wine tourism and services), the majority of papers are about motivations to visit wineries and the quality of services (Charters et al., 2009; Kolyesnikova & Dodd, 2009; Sekulić et al., 2017). This cluster falls outside the network, and can be identified as an old cluster, with the average year of publication being 2013.

7.4 Conclusions and Future Research

This aim of this work was to conduct a systematic review of wine tourism and customer behavior by means of a bibliometric analysis. It was implemented using a bibliographic coupling analysis and a content analysis that allowed us to determine the intellectual structure of wine tourism and consumer behavior and identify future lines of research and, in general, the footprint of the academy in this sector.

Our research shows the complexity of this field, which means that the relationship between wine tourism and customer behavior can be affected by many distinct factors. The study demonstrates that this topic dates back to the second half of 2002, which is consistent with extant literature that indicates that authors began to produce material on this topic in early 2002 (Gómez et al., 2019). Based on our research, we found and elucidated a number of avenues for future research that can help advance this field. As suggested by Lockshin et al. (1997) and Charters (2009), it would be useful to analyze consumer behavior and the segmentation of the wine tourism market in order that wineries can become more efficient when promoting their destinations in social networks or any other medium.

Our research revealed a distinct number and type of clusters on the topic studied and revealed it as an interesting field to delve into in order to improve the economics of wine companies and their corresponding regions. We suggest that future research could advance this specific literature by studying the relationship between producer branding and tourism destination branding, seeking to generate synergies between the two of them.

The comprehensive review conducted on wine tourism can guide scholars in planning their future research efforts. We found several lines of research that we expect to be developed in the coming years. First, regarding the wine tourism experience cluster, the cluster with the highest density, we expect that in the future it will be further developed and offer greater linkages with the rest of the clusters.

Some clusters require more attention. Specifically, the yellow cluster reveals that the vast majority of research focused on consumer loyalty has focused on analyzing how customer behavior relates to the tasting experience. However, future research could focus on analyzing the different profiles of wine tourists and forecasting their behavioral intentions.

Finally, we note that some trends highlighted in the literature on wine tourism, specifically wine tourism segmentation and its impact on the economy of the wineries and the region. Future research should focus on determining how the different segmentations can add value to both the winery and the region where the wine tourism activity takes place.

7.5 Practical Implications

This article provides useful and timely information on wine tourism for students, academics, winery managers, consultants and practitioners. The overview of the studies and clusters developed provides a solid research base for managers and graduate students in wine tourism evaluating research in this area. The findings of this bibliometric study may inspire future innovative research by master's or doctoral students interested in acquiring knowledge about lesser-known enotourism identified as receiving little attention in the literature. Ultimately, this article reveals useful new patterns and research opportunities for researchers and practitioners.

From an industry standpoint, the paper provides winery managers with information on the most prevalent consumer behavior patterns and management styles. In this way, our article can encourage wine tourism managers and professionals to reflect on their own management and thus explore different approaches when the situation calls for it. This article can help professionals to optimize winery resources, detect, praise and encourage preferred consumer behavior.

References

Asero, V., & Patti, S. (2011). Wine tourism experience and consumer behavior: The case of Sicily. *Tourism Analysis, 16*(4), 431–442.

Beames, G. (2003). The rock, the reef and the grape: The challenges of developing wine tourism in regional Australia. *Journal of Vacation Marketing, 9*(3), 205–212.

Björk, P., & Kauppinen-Räisänen, H. (2016). Local food: A source for destination attraction. *International Journal of Contemporary Hospitality Management, 28*(1), 177–194.

Brunner, T., & Siegrist, M. (2011). A consumer-oriented segmentation study in the Swiss wine market. *British Food Journal, 113*(3), 353–373.

Bruwer, J. (2002). The importance and role of the winery cellar door in the Australian wine industry: Some perspectives. *Australian and New Zealand Grapegrower and Winemaker, 463*, 96–99.

Bruwer, J., & Alant, K. (2009). The hedonic nature of wine tourism consumption: An experiential view. *International Journal of Wine Business Research, 21*(3), 235–257.
Bruwer, J., Coode, M., Saliba, A., & Herbst, F. (2013). Wine tourism experience effects of the tasting room on consumer brand loyalty. *Tourism Analysis, 18*(4), 399–414.
Charters, S. (2009). New world and mediterranean wine tourism: A comparative analysis. *Tourism: An International Interdisciplinary Journal, 57*(4), 369–379.
Charters, S., & Ali-Knight, J. (2002). Who is the wine tourist? *Tourism Management, 23*(3), 311–319.
Charters, S., Fountain, J., & Fish, N. (2009). "You felt like lingering..." Experiencing "real" service at the winery tasting room. *Journal of Travel Research, 48*(1), 122–134.
Chen, X., Bruwer, J., Cohen, J., & Goodman, S. (2016). A wine tourist behavior model for Australian winery cellar doors. *Tourism Analysis, 21*(1), 77–91.
Donthu, N., Kumar, S., Mukherjee, D., Pandey, N., & Lim, W. M. (2021a). How to conduct a bibliometric analysis: An overview and guidelines. *Journal of Business Research, 133*, 285–296.
Donthu, N., Kumar, S., & Pandey, N. (2021b). A retrospective evaluation of marketing intelligence and planning: 1983–2019. *Marketing Intelligence & Planning, 39*(1), 48–73.
Esau, D., & Senese, D. M. (2022). The sensory experience of wine tourism: Creating memorable associations with a wine destination. *Food Quality and Preference, 101*, 104635.
Fernandes, T., & Cruz, M. (2016). Dimensions and outcomes of experience quality in tourism: The case of Port wine cellars. *Journal of Retailing and Consumer Services, 31*, 371–379.
Festa, G., Shams, S. R., Metallo, G., & Cuomo, M. T. (2020). Opportunities and challenges in the contribution of wine routes to wine tourism in Italy–A stakeholders' perspective of development. *Tourism Management Perspectives, 33*, 100585.
Galloway, G., Mitchell, R., Getz, D., Crouch, G., & Ong, B. (2008). Sensation seeking and the prediction of attitudes and behaviours of wine tourists. *Tourism Management, 29*(5), 950–966.
Getz, D. (2000). *Explore wine tourism: Management, development & destinations*. Cognizant Communication Corporation.

Getz, D., & Brown, G. (2006). Critical success factors for wine tourism regions: A demand analysis. *Tourism Management, 27*(1), 146–158.

Gómez, M., Pratt, M. A., & Molina, A. (2019). Wine tourism research: A systematic review of 20 vintages from 1995 to 2014. *Current Issues in Tourism, 22*(18), 2211–2249.

Hall, C. M., Longo, A. M., Mitchell, R., & Johnson, G. (2009). Wine tourism in New Zealand. In *Wine tourism around the world* (pp. 150–174). Routledge.

Hall, C. M., Sharples, L., Cambourne, B., & Macionis, N. (2000). *Wine tourism around the world*. Butterworth-Heinemann.

Hernández, A. L., Alarcón, S., & Meraz Ruiz, L. (2022). Segmentation of wine tourism experience in Mexican wine regions using netnography. *International Journal of Wine Business Research, 34*(3), 427–446.

Hood, W. W., & Wilson, C. S. (2003). Informetric studies using databases: Opportunities and challenges. *Scientometrics, 58*, 587–608.

Kolyesnikova, N., & Dodd, T. H. (2009). There is no such thing as a free wine tasting: The effect of a tasting fee on obligation to buy. *Journal of Travel & Tourism Marketing, 26*(8), 806–819.

Lockshin, L. S., Spawton, A. L., & Macintosh, G. (1997). Using product, brand and purchasing involvement for retail segmentation. *Journal of Retailing and Consumer Services, 4*(3), 171–183.

López-Guzmán, T., Rodríguez-García, J., & Vieira-Rodríguez, A. (2012). Análisis diferenciado del perfil y de la motivación del turista nacional y extranjero en la ruta del vino del Marco de Jerez. *Gran Tour: Revista de Investigación Turística, 6*, 83–100.

Martins, J., Gonçalves, R., Branco, F., Barbosa, L., Melo, M., & Bessa, M. (2017). A multisensory virtual experience model for thematic tourism: A Port wine tourism application proposal. *Journal of Destination Marketing & Management, 6*(2), 103–109.

Roberts, L., & Sparks, B. (2006). Enhancing the wine tourism experience: The customers' viewpoint. In *Global wine tourism: Research, management and marketing* (pp. 47–55). Cabi.

Santos, V., Dias, A., Ramos, P., Madeira, A., & Sousa, B. (2021). Mapping the wine visit experience for tourist excitement and cultural experience. *Annals of Leisure Research*, 1–17.

Santos, V., Ramos, P., Almeida, N., & Santos-Pavón, E. (2020). Developing a wine experience scale: A new strategy to measure holistic behaviour of wine tourists. *Sustainability, 12*(19), 8055.

Sekulić, D., Petrović, A., & Dimitrijević, V. (2017). Who are wine tourists? An empirical investigation of segments in Serbian wine tourism. *Економика пољопривреде, 64*(4), 1571–1582.

Sigala, M. (2020). The transformational power of wine tourism experiences: The socio-cultural profile of wine tourism in South Australia. In *Social sustainability in the global wine industry: Concepts and cases*, 57–73.

Statista. (2023). *Number of visitors to wineries and wine museums in Spain.* https://www.statista.com/statistics/774012/monthly-number-of-visitors-to-the-wineries-and-museums-of-the-wine-routes-of-spain/

Szolnoki, G. (2018). New approach to segmenting tourists in a German wine region. *International Journal of Wine Business Research, 30*(2), 153–168.

Tafel, M., & Szolnoki, G. (2020). Estimating the economic impact of tourism in German wine regions. *International Journal of Tourism Research, 22*(6), 788–799.

Thach, E. C., & Olsen, J. E. (2003, July). Customer service training in winery tasting rooms: Perceptions of effectiveness by tasting room personnel. In *Third Australian Wine Marketing Conference*, Adelaide (pp. 26–27).

Thomas, B., Quintal, V., & Phau, I. (2013). Wine tourist profiling: Segmentation using enduring involvement. In *Proceedings of the International Management Development Association Conference* (pp. 124–126). National Taipei University.

Thomas, B., Quintal, V., & Phau, I. (2018). Wine tourist engagement with the winescape: Scale development and validation. *Journal of Hospitality & Tourism Research, 42*(5), 793–828.

Treloar, P., Hall, C. M., & Mitchell, R. (2004). Wine tourism and the Generation Y market: Any possibilities? *Paper presented at CAUTHE Conference*, Brisbane.

Van Eck, N. J., & Waltman, L. (2010). Software survey: VOSviewer, a computer program for bibliometric mapping". *Scientometrics, 84*(2), 523–538.

Verma, S., & Gustafsson, A. (2020). Investigating the emerging COVID-19 research trends in the field of business and management: A bibliometric analysis approach. *Journal of Business Research, 118*, 253–261.

Yuan, J., Morrison, A. M., Cai, L. A., & Linton, S. (2008). A model of wine tourist behaviour: A festival approach. *International Journal of Tourism Research, 10*(3), 207–219.

Zhang, Y., & Lee, H. (2022). Wine tourism experience effects on co-creation, perceived value and consumer behavior. *Ciência e Técnica Vitivinícola, 37*(2), 159–177.

Zupic, I., & Čater, T. (2015). Bibliometric methods in management and organization. *Organizational Research Methods, 18*(3), 429–472.

8

Wine Tourism in Israel: Preferences of Potential Wine Tourists

Eli Cohen

8.1 Wine Tourism

Global tourism has become one of the fastest-growing industries and wine tourism has become important and interesting for both academics and the wine industry in recent decades. Specifically, the trend of wine tourism is increasing in many countries with an increasing number of small wineries in many countries. The advantages of wine tourism for small wineries include an opportunity to increase sales, build brand loyalty, and educate consumers about wine (Dodd, 1995). Furthermore, wine tourism stimulates the economic development in the wine regions and increases regional employment and tourism (Carlsen, 2004; Sanders, 2004).

Wine tourism was initially focused on the visitation of wineries, vineyards, and wine cellars, with tasting, learning, and wine-purchasing

E. Cohen (✉)
Guilford Glazer School of Business and Management Department of Management, Ben-Gurion University of the Negev, Beer-Sheva, Israel
e-mail: elico@bgu.ac.il

J. Martínez-Falcó et al. (eds.), *Strategic Management in the Wine Tourism Industry*, https://doi.org/10.1007/978-3-031-54837-6_8

opportunities. Increasingly the traveler's broader interest in the entire wine destination experience, i.e., in exploring the complex wine-producing rural territory, its specific landscapes, people, history, and traditions, is recognized as a main driver for wine tourism (Holland et al., 2014). Charters and Ali-Knight (2002) provided an example of the Champagne region in France that has a wide "bundle of benefits" such as heritage buildings, historical sites, restaurants, and attractive landscapes as well as a range of wine tourism experiences that are offered by wine producers, such as visiting wineries, wine museums, and underground caves used to age the wine. This approach, sometimes referred to as "terroir tourism", acting beyond winery services, triggers several multiplier effects crucial to regional development (Kastenholz et al., 2022).

The combination of agriculture and tourism greatly impacts rural areas; they contribute to preserving culture and tradition and improving the local economy. Tew and Barbieri (2012) showed that agritourism mostly serves to capture new farm customers, educate the public about agriculture, and enhance the quality of life for the farm family, which represents both economic and non-economic benefits. In recent years there has been an increase in the number of visitors who are looking for experiences in which they can explore and enjoy nature and learn from and about it. People are looking to visit rural places, and farms and meet and work with farmers, basically, everything that agritourism offers (Joyner et al., 2018). This type of activity is particularly attractive to the urban population looking for an interest in connecting to nature and learning about the agricultural life of the inhabitants of rural areas (Lupi et al., 2017). The tourism industry is based on many small organizations with a limited and overlapping market share, which leads to the necessity of using marketing strategies in this industry (Kiralova & Pavliceka, 2015). In addition, individual businesses depend on each other; therefore, to survive, they must cooperate.

8.2 Wine Tourism in Israel

Wine has a long history in Israel, with wine produced in the area dating back thousands of years to biblical times. However, the modern wine industry in Israel began to develop in 1882 when James Rothschild funded the planting of grapes and the building of a small winery in Rishon Le-Zion, in the central region of the country. The idea was to develop the wine industry in the Holy Land and to produce and sell wine, mainly for religious purposes. Today, the number of wineries operating in Israel is estimated at about 350, most of them small or family wineries (so-called boutique wineries). Despite this significant figure, the Israeli wine market is still dominated by about 10 large commercial wineries, which produce about 90% of Israeli wines. The main wine regions are in the north of the country and in the mountains around Jerusalem. However, in recent years, wineries have also been established in the south of the country in desert areas.

Israel has a growing wine tourism industry, with several regions known for producing high-quality wines. Some popular wine regions in Israel include the Galilee, Golan Heights, and the Judean Hills. Visitors can participate in wine tastings and tours at many of the wineries in these regions, as well as learn about the history of wine-making in Israel. The Negev in Israel is an emerging wine region that has the potential for differentiation and recognizes as a wine region due to the severe stressful conditions of the vine and the crops in the desert. Simultaneously with the development of the Israeli wine industry, the desire of wine lovers to experiment with new wines, new varieties, and special wines has also developed. Such wines are produced by large wineries and can be purchased in wine shops, but wines from small wineries can be found mainly in the winery that produces the wine or in special wine stores, which do not necessarily carry wines from many small wineries. It was found that wine tasting is the main motivation for visiting wine tourism sites and wineries.

About two decades ago, wine tourists in Israel mainly focused on getting information about the winery, the production processes, and wine tastings, and even purchasing wine at the wine shop in the winery or at the visitor center. Jaffe and Pasternak (2004) were the first scholars who suggested developing wine trails in Israel as tourist attractions. Their study shows that nearly half the winery visitors interviewed visited a winery for the first time. They concluded that some of the marketing strategies can be implemented by individual wineries while others can be brought about only by the cooperative effort of several wineries.

In areas where there are many wineries, such as in the north of the country and in the center, there are many hotels, guest rooms, and restaurants within a short drive from the winery, so wine tourists can dine and even spend the night in the area after drinking alcohol. Since public transportation between wineries is almost non-existent, the only way to travel to the wineries is by private car or organized transportation. In Israel, going on spontaneous tours using a private vehicle is usually customary. Organized tours in wineries were of less importance for the Israel potential wine tourists (Cohen & Ben-Nun, 2009). However, there are wineries in Israel that offer guided tours of vineyards and facilities, which can include tastings of their wines and an understanding of the wine-making process, the winery's history, and the region's terroir.

8.3 Preferences of Potential Wine Tourists

An exploratory study has been conducted in Israel between December 2006 and February 2007 and analyzed by Cohen and Ben-Nun (2009). Participants (over the age of 18 years) were asked to rate each attribute's importance in terms of their own perceptions of each of 42 features (on a Likert-type 1–5 scale) with respect to their decision to visit wineries or wine regions. Out of 372 respondents, 86 (24.4%) were between 18 and 24, 125 (35.4%) between 25 and 34, 58 (16.4%) between 35 and 44, 45 (12.8%) between 45 and 54, and 39 (11.0%) over the age of 54. The subjects live in different regions of Israel, 58 respondents live in Jerusalem (15.6%), 43 in Tel Aviv (11.5%), 43 Beer-Sheva (11.5%), and 21 in Haifa (5.6%). All other respondents live in 120 different locations in Israel. Their results show that the most important feature for the respondents was "wine tasting offered by the winery" followed by "receiving information about the wine-making process". The special atmosphere in the winery was also important for the potential visitors.

Features like children's activities in the area, sports facilities, and shops offering special arts or handicrafts were the least important. The authors utilize Principal Component Analysis (PCA) to derive three factors, representing core features, as follows: (1) "winery atmosphere"; (2) "cultural activities"; and (3) "family activities". These results indicate that the motivations for choosing a wine region as a tourist destination go beyond a simple desire to experience wine. A tourist's decision to visit a specific winery is also shaped by the willingness to experience the region's features and to participate in other activities or attractions during the visit. However, although the "family activities" was of less importance for the subjects, it was still important for families, mainly for young people with young kids.

Another study by Ben-Nun and Cohen (2009) utilized the Best-Worst Scaling (BWS) methodology to measure the relative importance of the wine tourism features of potential wine tourists in Israel and explore possible latent clusters among the respondents. The BWS approach, also known as maximum difference scaling, was first published in 1992 (Finn & Louviere, 1992). The attributes are presented in various combinations, based on experimental designs, and the subjects are forced to make trade-offs between the items and to choose the most preferred item (called "most" or "best") and the least preferred item ("least" or "worst") in a set of items, called "choice sets". As respondents can only choose one "most" and one "least" preferred item in each choice set, they are necessarily required to make trade-offs between items or benefits (Cohen, 2003). The method provides better discrimination between the attributes, and it helps to overcome many of the limitations of scale-based surveys such as Likert-type scales (see Crask & Fox, 1987, Cohen, 2009, Cohen & Lockshin, 2016, Hein et al., 2008, Lockshin & Cohen, 2011, 2015, Marley & Louviere, 2005). Ben-Nun and Cohen (2009) study which took place in Israel at the outset of 2008, utilized 7 important attributes which represent wine tourists' expectations in visiting wineries in wine regions. The authors did not find significant differences between segments based on gender, age, frequency of drinking wine, and frequency of visiting wineries, regarding the rating of the features (based on 265 adults' responses). However, using the involvement in wine as a segment base showed significant differences among the different segments regarding the rating of the features.

Consumer involvement has been widely used in marketing studies. Involvement is considered as a motivational force acting on consumers' decision processes (Lockshin et al., 2006), which might explain various behavioral outcomes, such as product knowledge, frequency of purchase, and money spent. Several scholars have asserted the importance of involvement in the wine tourism experience and considered it to be a good predictor of wine tourism behavior (Gross & Brown, 2006; Lockshin & Spawton, 2001; Sparks, 2007). Respondents were asked to answer three questions that described their wine-involvement behavior patterns. Based on the average score of the three involvement questions (1–5 Likert-type scale), respondents were classified into three categories: "high-involved", "medium-involved", and "low-involved". Ben-Nun and Cohen (2009) reported that the low-involved subjects preferred the possibility of participating in a range of activities within the region, while the highly involved subjects gave a lower importance level for this feature. The highly involved segment showed great interest in wine tourism activities and was more interested in getting information about the winery and the wine region. "Activities related to the wine region" was the most preferred activity and "guided tours" were of the least importance to the potential wine tourists in the study. Further analyses showed two latent clusters that represent the different cores of attributes: Cluster 1 consists of about 70% of the respondents and is labeled as "activities in the region" since its subjects are mainly focused on outdoor activities. Cluster 2 is labeled as "activities in the winery" and consists of winery-related features such as staff knowledge and guided tours in the winery.

A recent study aims to test the potential of the Negev, a southern region in Israel, to attract potential wine tourists and to explore the preferences of the potential tourists when they come to visit wineries in the Negev. The Negev has the potential for differentiation and recognizes as a wine region due to the severe stressful conditions of the vine and the crops in the desert. The Negev has spectacular views, an ancient tradition of growing vines and wine production, and innovative and leading technologies. The area is rich in wineries and vineyards proving that quality wine can be produced in the challenging conditions of the desert (Gaudreau, 2013). The main tourism activities in the Negev focus on small businesses such as artisans and crafts (souvenirs, galleries), extreme activities, and visiting farms, beehives, and ancient agriculture. The great advantage is expressed when the experience of the visitors is inclusive and contains several activities in several different places such as tasting food, visiting wineries, arriving at historical and cultural buildings, visiting and working on farms, and more. These actions create an overall experience that adds value to local products. The development of the wine tourism network showed that an important component in differentiating the Negev as a wine region is the cooperation between the wineries and the other tourism businesses in the Negev. All of these make the Negev a unique and attractive tourist destination, with economic and social opportunities.

The study was conducted in Israel for adults above 18 years old. A questionnaire was distributed among Israeli citizens, over 60% of them live in the central part of Israel (more than 100 km north of Negev). The survey was published for groups of wine lovers, and different age groups (students, family, groups of friends, groups of parents' friends, etc.). The data were collected in May 2022, and 114 questionnaires were valid for the analyses. Most of the respondents to the questionnaire belonged to the ages 18–29 (53%), 43% of the subjects were women and the rest were men. Also, about 70% of those answering the questionnaire live in the central part of the country (about 100–150 km from the north Negev), and most of them live in urban municipalities. Only about 26 participants indicated that they had visited wineries in the Negev in the past. The BWS scaling method was implemented as the method allows evaluation of the relative importance of each attribute compared to the most important attribute derived from the survey (Cohen, 2009).

The most important feature for the respondents was "Visit the winery and cellar door", and hence it was defined as 100% important. The following feature was "purchasing local agricultural products" and "cultural events in the region". The least important feature is "children's activities" (see Fig. 8.1). There were no significant differences between "visit the winery and cellar door", "purchasing local agricultural products", and "cultural events in the region". This means that visitors to a wine region are looking for all three features with a preference for a winery and cellar door visit, but they also expect other features. Wine tourists do not visit a wine region just for buying local agricultural products, but they are expecting to buy such products during their visit to a wine region. Children's activities are the least important in this study, probably because most of the respondents in this study (about 70%) are without children and/or subjects who have children are not willing to visit wineries with them.

Other features that were assessed were (in order of importance) "a website for pre-booking", "suggestion of winery routes", "packages of accommodation", "other activities", and "transportation between accommodation and wineries". Unsurprisingly, transportation between accommodations and wineries is unimportant as most visitors use their own vehicles to visit the Negev and for vacation purposes. A minority of visitors use organized tours to the Negev attractions. It is not surprising that

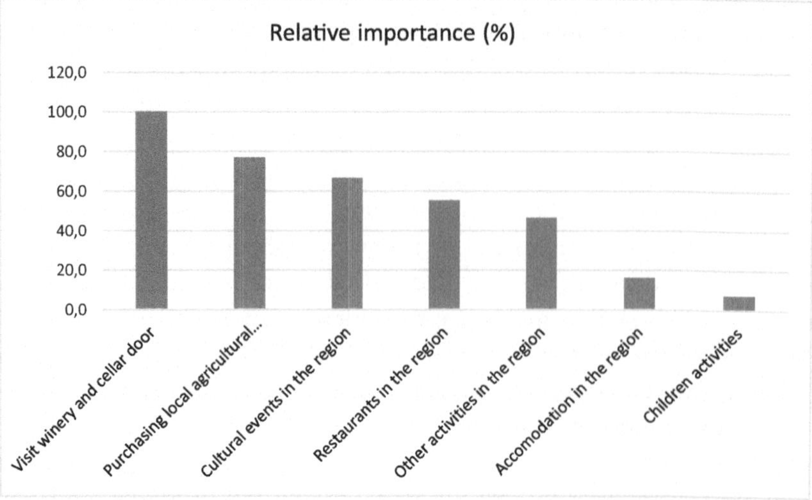

Fig. 8.1 Relative importance of wine tourism features (compared to "visit a winery and cellar door")

potential visitors are willing to pre-design the visits, as most of them are young and familiar with using the web.

This study is considered exploratory due to the small number of respondents. A broader survey of more representative samples should be conducted to better understand how to construct preferred wine tourism packages in the Negev according to wine tourists' preferences. However, the strategic plan should be attractive to potential customers, not only wine lovers, in a dynamic and customized way. The tourism package will allow for an attractive and temporary product, which will change according to the needs of both the customers and the businesses. An important tool to attract visitors is a marketing plan to increase the number of visitors and raise awareness of the area, the businesses, and the wineries. The marketing strategies that have been found to lead to maximizing profits in tourist sites include focusing on digital media and building a visual and accessible brand for the destination (Buhalis, 2000). Today's common marketing platform for tourism businesses is Instagram and other social media applications. This function makes it possible to increase the experience of visiting the existing tourist sites while marketing the destination to the visitors themselves.

8.4 Conclusions

Visitors are looking for experiences in which they can explore and enjoy nature and learn from and about it. However, the wine tourism experience is often part of an overall "bundle of benefits", which includes a winery visit and/or wine tasting and other activities such as visiting the area, tasting food, enjoying the scenery, and participating in other activities and local attractions. These actions create an overall experience that adds value to local products and businesses. The evolution of the wine tourism network plays a pivotal role in distinguishing a wine region, underscored by the collaboration efforts between wineries and various tourism enterprises. Israeli tourists, particularly the younger demographic, have increasingly shown enthusiasm for wine and wine-related experiences. This burgeoning interest indicates a promising route for wine tourism in Israel, necessitating the expansion of tour packages across wine regions. Such initiatives, supported by local authorities, enhance visitor experiences and stimulate economic growth for regional businesses.

To increase awareness of the special wines produced by boutique wineries, and therefore to attract visiting boutique wineries, it is recommended to hold wine events in the central and the northern part of the country, where wine-makers from the wineries will attend to tell the story of the winery and offer wine tastings and sell their wines at attractive prices. It has been found in many studies that information transmitted by word of mouth is critical in the field of tourism, where the products can be evaluated mainly according to the consumption indicators of the customers (Jalilvand & Samiei, 2012). As Levy and Cohen (2016) showed in their study, organizing such evenings and the presence of the wine-makers during such events are an important and effective marketing tool, especially for small wineries, and will allow the winery owners to reach additional audiences. The existence of wine evenings can lead to a marketing advantage as in this way the winery owners reveal their wines to people who would not necessarily come directly to the winery. In addition, since these might be intimate events, this is a good opportunity to expand the number of wine lovers who will explore wine from different wineries and regions, and the wineries will create new business opportunities.

To provide local agricultural products, it is recommended to establish stands for local agricultural produce in wineries and various wine region tourism businesses. The combination of businesses in the field of wine and businesses in the field of rural tourism is known to enable the creation of innovative collaborations and the creation of competitive advantage so that the partners can work toward achieving common goals and objectives (Karampela et al., 2017). In addition, the possibility of buying local agricultural produce is found to be one of the most significant variables for tourists when visiting the winery, so the establishment of these stations will allow increased exposure to agricultural products produced in the region and in particular wines from special wine regions such as the Negev.

A major collaboration we recommend is the creation of joint tourism packages that include winery or wineries visits, and cultural events in the region. As Del Chiappa et al. (2022) stated that most important aspects that make wineries attractive are offering wine tasting and "tours and visits" to provide visitors with wine specialists and to make the surrounding area and natural environment as pleasant as possible.

As most studies on wine tourism focus on tourists who have visited wineries, the studies in Israel focus on potential visitors and introduce a framework for future studies that can focus on the perception of features during and post-visit of visitors. Israel, as an emerging wine market with an increasing number of wineries, the Israeli wineries with the collaboration of tourist attractions in the region might benefit greatly from understanding wine tourism features that most influence potential visitors.

References

Ben-Nun, L., & Cohen, E. (2009). *Using best-worst scaling method to classify potential wine tourists*. Presented at Bacchus Goes Green, 4th Interdisciplinary and International Wine Conference, Dijon, Bourgogne, France, July, 8–10.

Buhalis, D. (2000). Marketing the competitive destination of the future. *Tourism Management, 21*(1), 97–116.

Carlsen, J. (2004). A review of global wine tourism research. *Journal of Wine Research, 15*(1), 5–13.

Charters, S., & Ali-Knight, J. (2002). Who is the wine tourist? *Tourism Management, 23*, 311–319.

Cohen, E. (2009). Applying best-worst scaling to wine marketing. *International Journal of Wine Business Research, 21*(1), 8–23.

Cohen, S. H. (2003), Maximum difference scaling: Improved measures of importance and preference for segmentation. In *Sawtooth Software Conference Proceedings*, Sequim, WA.

Cohen, E., & Ben-Nun, L. (2009). The important dimensions of wine tourism experience from potential visitors' perception. *Tourism and Hospitality Research, 9*(1), 20–31.

Cohen, E., & Lockshin, L. (2016). Best-worst scaling method: Application to hospitality and tourism research. In M. Uysal, Z. Schwartz, & E. Sirakaya-Turk (Eds.), *Management science in hospitality and tourism: Theory, practice and applications*. Apple Academic Press.

Crask, M. R., & Fox, R. J. (1987). An exploration of the interval properties of three commonly used marketing research studies: A magnitude estimation approach. *Journal of the Marketing Research Society, 29*(3), 317–339.

Del Chiappa, G., Martin, J. C., & Roman, C. (2022). Developing wine tourism experience. A discrete choice analysis using best-worst scaling data. *Wine Economic and Policy, 11*(1), 107–126.

Dodd, T. (1995). Opportunities and pitfalls of tourism in developing wine industry. *International Journal of Wine Marketing, 7*(1), 5–16.

Finn, A., & Louviere, J. J. (1992). Determining the appropriate response to evidence of public concerns: The case of food safety. *Journal of Public Policy and Marketing, 11*(1), 12–25.

Gaudreau, K. (2013). The grapes of math: Advancing the science of wine production in Israel's Negev Desert. *Alternatives Journal, 39*(4), 40.

Gross, M. J., & Brown, G. (2006). Tourism experiences in a lifestyle destination setting: The roles of involvement and place attachment. *Journal of Business Research., 59*(6), 696–700.

Hein, K. A., Jaeger, S. R., Carr, B. T., & Delahunty, C. M. (2008). Comparison of five common acceptance and preference methods. *Food Quality and Preferences, 19*, 651–661.

Holland, T., Smit, B., & Jones, G. V. (2014), Toward a conceptual framework of terroir tourism: A case study of the Prince Edward County, Ontario wine region, *Tourism Planning and Development, 11*(3), 275–291. https://doi.org/10.1080/21568316.2014.890125.

Jaffe, E., & Pasternak, H. (2004). Developing wine trails as tourist attraction in Israel. *International Journal of Tourism Research, 6*(4), 237–249.

Jalilvand, M. R., & Samiei, N. (2012). The impact of electronic word of mouth on a tourism destination choice. *Internet Research, 22*(5), 591–612.

Joyner, L., Kline, C., Oliver, J., & Kariko, D. (2018). Exploring emotional response to images used in agritourism destination marketing. *Journal of Destination Marketing & Management, 9*, 44–55.

Kastenholz, E., Paco, A., & Nave, A. (2022). Wine tourism in rural areas—Hopes and fears amongst local residents. *Worldwide Hospitality and Tourism Themes, 15*(1), 29–40.

Karampela, S., Kavroudakis, D., & Kizos, T. (2017). Agritourism networks: Empirical evidence from two case studies in Greece. *Current Issues in Tourism, 22*(12), 1460–1479.

Kiralova, A., & Pavliceka, A. (2015). Development of social media strategies in tourism destinations. *Procedia-Social and Behavioral Sciences, 175*, 358–366.

Levy, T., & Cohen, E. (2016). Success marketing factors for boutique wineries: Perception of wine store managers. In *Proceedings of the 9th International Conference of the Academy of Wine Business Research* (AWBR), UniSA, Adelaide, South Australia, February 16–19 (pp. 283–291). http://academyofwinebusiness.com/wp-content/uploads/2016/03/30.-Success-marketing-factors-for-boutique-wineries-Perception-of-wine-store-managers.pdf

Lockshin, L., & Cohen, E. (2015). How consumers choose wine—Using best worst scaling across countries, Chapter 8. In J. J. Louviere, T. N. Flynn & A. A. J. Marely (Eds.), *Best-worst scaling, theory, methods and applications*. Cambridge University Press.

Lockshin, L., & Cohen, E. (2011). Using product and retail choice attributes for cross-national segmentation. *European Journal of Marketing, 45*(7/8), 1236–1252.

Lockshin, L., Jarvis, W., d'Hauteville, F., & Perrouty, J. P. (2006). Using simulations from discrete choice experiments to measure consumer sensitivity to brand, region, price, and awards in wine choice. *Food Quality and Preference, 7*, 166–178.

Lockshin, L., & Spawton, T. (2001). Using involvement and brand equity to develop a wine tourism strategy. *International Journal of Wine Marketing, 13*(1), 72–81.

Lupi, C., Giaccio, V., Mastronardi, L., Giannelli, A., & Scardera, A. (2017). Exploring the features of agritourism and its contribution to rural development in Italy. *Land Use Policy, 64*, 383–390.

Marley, A. A. J., & Louviere, J. J. (2005). Some probabilistic models of best, worst, and best-worst choices. *Journal of Mathematical Psychology, 49*, 464–480.

Sanders, D. (2004). Wine tourism and regional development: A case study of the Margaret River Region, Western Australia. In *Proceedings of the International Wine Tourism Conference Margaret River Western Australia*, May 2004.

Sparks, B. (2007). Planning a wine tourism vacation? Factors that help to predict tourist behavioral intentions. *Tourism Management, 28*(5), 1180–1192.

Tew, C., & Barbieri, C. (2012). The perceived benefits of agritourism: The provider's perspective. *Tourism Management, 33*(1), 215–224.

Part IV

New Management Tools for the Wine Tourism Sector

9

Forecasting Wine Tourism: A Survey of Methods

Martin Kunc

9.1 Introduction

Wine tourism can be defined as "visitation to vineyards, wineries, wine festivals, and wine shows for which grape wine tasting and/or experiencing the attributes of the grape wine region are the prime motivating factors for visitors" (Hall & Macionis, 1998, p. 197). Wine tourism is recognised as an important area of tourism in many countries around the word, especially wine-producing countries, because it can provide economic, social and environmentally sustainable tourism development in rural areas (Alonso & Liu, 2012). In a recent survey, wine tourism was found to be a key element for the sustainable development of wine regions (Sustainable Wine Tourism: A Global Survey, 2022). The most common resources and capabilities for developing wine tourism include incorporating wine tastings, services and wine sales into winery visits, vineyard and wine cellar tours and wine festivals (Torres et al., 2021).

M. Kunc (✉)
University of Southampton, Southampton, UK
e-mail: M.H.Kunc@soton.ac.uk

J. Martínez-Falcó et al. (eds.), *Strategic Management in the Wine Tourism Industry*,
https://doi.org/10.1007/978-3-031-54837-6_9

These resources generate short-term revenue not only for wine firms but also for tour operators that offer a variety of complementary services related to the wine experience such as dinners and excursions (Torres et al., 2021).

There are different ways to quantify wine tourism in terms of number of visitors (Mitchell & Hall, 2006), but there is no agreement on a specific method. Forecasting represents a different challenge than capturing actual consumer behaviour because forecasting implies a development path of future consumer behaviour. Steinhagen et al. (1998) provided an exploratory study of forecasting models in the wine industry where they found wide use of judgmental forecasting methods like aggregating sales forecasts provided by agents or using historical time series to extract a trend. However, there is no clear information in terms of forecasting wine tourism.

Forecasting wine tourism is critical to achieve sustainability because it impacts across multiple areas: winery, traffic, natural areas and whole supply chain. For example, the capacity of the winery to receive tourists is critical for the activity and any expansion of buildings will affect the sustainability of the winery. Traffic can be associated with cars from tourists driving around the countryside and generating pollution. Natural areas can suffer if the wine tourists are only focused on the wineries and drinking wine rather than protecting and enjoying the natural area where the winery is located. Finally, the emissions for sending wines to consumers may be reduced if consumers buy their wine locally and in bulk quantities during winery visitation. Forecasting can support investments and decisions to facilitate sustainable progress in rural areas. However, the adoption of forecasting in wine tourism is still in its infancy.

This chapter will explore the state of the field in terms of forecasting, its use in wine and wine tourism. Then, a framework to support forecasting in wine tourism is proposed at the end of the chapter.

9.2 Literature Review

Methods to Predict or Forecast

Kunc (2018) proposed a classification of different methods to predict/forecast future values of variables. See Table 9.1 for a list of the methods. The first column indicates the name of the method followed by a brief description. The second column discuss the acceptability among decision-makers. Third column shows the level of skills required to apply the method. The final column suggests the type of uncertainty for the method indicated.

To summarise, there are multiple methods using qualitative and quantitative data and diverse levels of involvement from the decision-makers. All methods have strengths and weaknesses in terms of skills, data required and acceptability. However, the critical aspect to consider is the type of uncertainty being addressed with the method. For example, wine production is usually forecasted using econometric models and trend models because their data has been collected for many years because wineries must provide this information to governments. In the case of wine tourism, no data is captured except by some regional authorities responsible for wine routes or local association of wineries, but the reporting is ad hoc.

Forecasting Tourism

A synopsis of wine tourism literature was conducted by Mitchell and Hall (2006) who identified seven themes such as the wine tourism product and its development; wine tourism and regional development; the size of the winery visitation market; winery visitor segments; the behaviour of the winery visitor; the nature of the visitor experience; and biosecurity. While there are many studies that tried to quantify wine visitation in terms of number of visitors, there is no agreement on a specific method (Mitchell & Hall, 2006). Some issues are the lack of accurate historical data, multiple dimensions to use for forecasting and lack of effective methods.

Table 9.1 Methods for forecasting (adapted from Kunc, 2018)

Methods	Acceptability	Level of skills required	Type of uncertainty
Econometric models They are simultaneous sets of multiple regression equations	High due to the importance of economic theory in the management of public policies	This tool is basically employed by economists to forecast future developments in the economy based on past relationships among economic variables. Consequently, it requires knowledge of economic theory	Predictable future
Simple and multiple regression models They explain variations in dependent variables due to the impact of one or more independent variables	High due to the availability of software and the widespread existing knowledge about them	Basic regression analysis can be performed by any graduate with training in statistics. Sophisticated regression analyses require specific training	Predictable and alternative futures
Trend analysis These models assume the future will result from the existence of a long-term trend	High due to relatively "safe", and extensively accepted, assumption that the future depends on the past. Additionally, this analysis can also be performed using widely available spreadsheet software	The level of skills depends on the type of technique to be performed since there are numerous techniques, e.g. simple average, exponential smoothing, logarithmic, with diverse levels of complexity. Additionally, complexity is generated by the identification of changes that may affect the trends	Predictable future

(continued)

Table 9.1 (continued)

Methods	Acceptability	Level of skills required	Type of uncertainty
Simulation Simulation are mathematical models employed for understanding current operations and forecasting future performance	Middle because some models can be very complex and difficult to understand (black box). On the other hand, it is widely accepted due to its flexibility to represent any system	There are three main types of simulation for business: system dynamics, discrete event simulation and agent-based simulation. Given the need for long-term perspectives, system dynamics and agent-based simulation are the two methods usually employed. It requires good expertise to develop the models	Predictable alternative, and a range of futures
Operational and executive estimates These estimates involve aggregating the opinions of diverse people	High due to extensively accepted assumption that experience of experts will be able to identify situations where historical data is not available	The skills required will be simple if the number of estimates are low, e.g. a simple average. However, the complexity may increase if the process involves many estimates	Predictable, alternative and a range of futures
Customer surveys/ market research They are primary data collection methods involving questionnaires, sampling, and statistical analysis	High as every business needs to be informed about customers' motivations or behaviours	The skills required are medium since it requires knowledge about questionnaire design, statistical sampling and analysis even if the research is performed externally	Predictable, alternative and a range of futures

(continued)

Table 9.1 (continued)

Methods	Acceptability	Level of skills required	Type of uncertainty
Scenario It is a systematic method for thinking about the future in a consistent and plausible manner	Medium as it is based in qualitative perceptions without quantitative data to validate them	There are simple methods to develop scenarios. However, the difficulty is in the acceptance of the results due to its long-term perspective	A range of futures
Delphi Method This method consists of the development of a consensus across a set of experts about the future	Medium due to the complexity of the method in terms of access to experts, questionnaire design and process	Simple after learning the method. However, it needs a systematic application as well as identifying and accessing the experts	A range of futures and true ambiguity

Literature related to forecasting tourism focuses on the tourist arrivals into the country or region (Jiao & Chen, 2019; Song & Li, 2008). Most studies use tourist arrivals as dependent variable even though time series are older than 10 years. One of the limitations of tourism forecasting is their view of tourism demand at a broad perspective, e.g. tourist arrivals, rather than visitors arriving at a specific tourist attraction, e.g. a museum, city, winery, etc. If they focus on visitors arriving at a specific attraction, it will help the management of an attraction for better planning rather than simply guessing the number of people.

9.3 Methodology

A systematic literature review is a synthesis of evidence within a particular domain by using a systematic method to search and categorise the literature (Connolly et al., 2012). The method for a systematic literature review has three stages: the selection of journals, the selection of keywords for search and the selection of relevant articles followed by their analysis (Connolly et al, 2012).

In this research, articles on wine tourism were identified and gathered using SCOPUS (Connolly et al., 2012). Articles were identified by searching the keywords and abstract for the search terms. Firstly, I employed "Forecast* AND Wine" in only BUSINESS category in Scopus. The total number of papers found using these keywords were 39. Then all abstracts were read, and 15 papers were selected. Secondly, I employed "Scena* AND Wine" in only BUSINESS category in Scopus. The total number of papers found using these keywords were 86. Then all abstracts were read, and 21 papers were selected. The papers selected include forecasting and scenarios for wine and tourism. The objective is to compare both practices to uncover different practices. All papers reviewed are numbered and included in the appendix. Table 9.2 provides broad statistics about the papers included. The number of papers has increased over time, but the trend is not stable. From the selection, most of the papers are related to forecasting and scenarios associated with production and sales rather than wine tourism. Consequently, the journals with higher number of papers are production or non-tourism journals. Interestingly, forecasting and scenarios are not even key topics in wine journals with only 10 papers for both wine journals over 14 years. Finally, there is a clear strength in Italian research on forecasting and scenarios followed by other traditional wine-producing countries. This result raises an interesting question for research regarding the interrelationships between Italian academics and the wine industry.

Table 9.3 displays the number of papers associated with each of the methods presented in the literature review (see Table 9.1). The result shown in Table 9.3 is understandable given the existence of data related to sales and production, as well as the interest of economists on forecasting with 11 paper using econometric models. There is some interesting work on addressing lack of historical data by using scenarios (10 papers) and simulations (9)

Another search using both keywords, forecast and scena, and tourism provided 1094 and 1096 papers, respectively. Moreover, a search using

Table 9.2 Descriptive information from selection of papers

Concept	Results										
Number of papers per year	2009	2011	2012	2013	2017	2018	2019	2020	2021	2022	2023
Distribution between tourism and production/	1	1	1	1	6	3	4	7	3	5	2
sales	*Production/sales*				30						
	papers:				4						
	Tourism papers:				34						
	Total papers:										
Distribution by journal	Journal of				9						
	Cleaner				5						
	Production				5						
	Journal of Wine				11						
	Economics				4						
	International										
	Journal of Wine										
	Business										
	Research										
	Other										
	non-tourism										
	Other tourism										
Geographical distribution of authors in	Italy				10						
number of papers	Australia				5						
	Spain				4						
	USA				3						
	France				2						
	Chile				2						
	Other countries				9						

Table 9.3 Usage of methods

Methods	Number of papers[a]
Econometric models	11
Simple and multiple regression models	2
Trend analysis	0
Simulation	9
Operational and executive estimates	2
Customer surveys/market research	2
Scenario	10
Delphi Method	0
Other methods	2

[a]The number of papers does not match because some papers use more than one method

the keyword "wine tourism" provided 341 articles. Both searches provide a clear idea of the relative limited practice of forecasting and scenarios within wine researchers.

9.4 Findings

Since this paper discusses only wine tourism forecasting, the findings will only cover the papers number [11], [13], [28] and [34].

Distribution of Methods Employed

Three papers [11, 13] use scenarios, one paper employs estimates [28] and another paper presents a simulation study [34]. The authors of paper [11] employ experts' analysis to define scenarios for different strategic wine initiatives. Study [13] uses experts' analysis to create scenarios about the future of gastronomic festivals and events. The study in [29] uses existing literature on the effects of climate change on wine tourism in the Northern New York, historical information and seven qualitative interviews with local and regional wineries. Based on this information, the authors provide an outlook for winc tourism. Paper [34] uses survey and simulation to forecast the development of wine tourism.

Context of the Application

Paper [11] applies to the Chinese market with detailed discussion of three regions: Shandong, Ningxia and Hebei. Study [13] discusses scenarios for gastronomic festivals and event globally. In study [29], the application is wine tourism on the Northern area of New York state. Study [34] is performed in Chile considering all wine regions as destinations.

Data Employed

Paper [11] employs existing academic research and experts' knowledge. Study [13] uses experts' insights to create scenarios. Study [29] uses literature and interviews to create estimates on the evolution of wine tourism in Northern New York. In paper [34], the author employs a survey to potential wine tourists to identify their preferences as wine tourists, which are then incorporated into a system dynamics model to simulate the different trajectories for wine tourism.

Time Horizon of the Study

There is no clear time horizon in papers [11] and [13], but the implications can be considered for less than 5 years. Study [29] looks at data up to 30 years old in terms of temperature and rain regimes to determine the impact of climate change. The forecast performed in paper [34] covers tourism over a year cycle, 12 months.

Key Insights Generated

Paper [11] suggests the implementation of strategic plans in three regions to attract not only the emerging domestic wine tourism industry but also international tourists. The plans involve the development of new wineries with tourist-friendly facilities, implementation of wine routes and wine-themed hotels. For one of the regions, its location close to major urban centres is perfect to grow it further in the next years. In another region, the focus is on the development of high-quality small boutique wine producers emphasising sustainability and green practices. In this case, tourists can be attracted through tour packages and facilitating travel planning. The final region is close to Beijing and host of an international wine festival. With different climates and terroirs, as well as authentic Chinese food culture, this region is ready to grow if there is more support to promote it. In all regions, the authors suggest focusing on the Millennial generation.

Study [13] identifies three economic trends which will affect the future of gastronomic festivals, wealth, technology and resources, leading to an increase of exclusive gastronomic events underpinned by best wines, food and finest accommodations. A countertrend can be environmental issues that will dissuade attendees from travelling to specific festivals.

In study [28], they suggest the outlook for wine tourism is mixed since it has been growing historically due to the attractiveness of the area as popular vacation destination, as well as the qualities of grapes due to higher temperatures, but there is increasing competition from new wineries, breweries and distilleries together with a strong seasonality. There is also a set of strategic recommendations to address the outlook.

Paper [29] demonstrates the importance of generating and maintaining a presence in the tourists' mind to grow wine tourism over time. One of the main issues of wine tourism is the experiential nature associated with wine tasting. This experience is difficult to keep active over time to facilitate word of mouth and the adoption of wine tourism by other potential tourists. The author argues that wineries should use newsletters with promotions and events aimed at previous visitors to keep their experience alive. This strategy can be complemented with merchandise for the wine tourists such as glasses with the winery logo and special bottles.

9.5 Discussion

The Role of Forecasting and Scenarios in Wine Tourism

Unfortunately, the systematic review of the literature does not present a good situation for the role of methods to evaluate the future of wine tourism. The evidence is very limited with few articles mostly describing current situations and based on authors' expertise providing scenarios or recommendations for strategic plans. This situation may be originated by the lack of extensive data evaluating the demand of wine tourism as well as arrivals of tourists to wineries, wine routes, wine regions and wine festivals.

Clearly, there is a need to start collecting more data about it. The use of apps and big data may be a suitable source to discover more about wine tourism and help regional authorities, wine managers and industry associations to plan its development in a sustainable way.

Proposed Framework

Three steps comprise this framework: a survey to obtain the data for the forecasting model, a forecasting model based on simulation and scenario development to test different futures. The framework is based on Kunc (2009) and Kunc and O'Brien (2017).

Step 1—Survey the wine tourist

Mitchell and Hall (2006) describe typical winery visitors as 30–50 years old, well educated, professional and in the moderate to high-income segment. While this is a clear segment, the main question is to find the closest market (in terms of distance) where this segment is present. Therefore, the forecasting model will provide the potential market size for the type of winery visitor described by Mitchell and Hall (2006). Additional data needs to be captured to measure the effect of word of mouth and advertising campaigns on the intention of making a future visit to a winery (Kunc, 2009). The effect of word of mouth is included as coefficient of imitation and the effect of advertising campaigns is represented as coefficient of innovation in the forecasting model. Another question to ask is how many times they expect to have the wine tourism experience, which is denoted as coefficient of forgetting in the model.

Step 2—Forecasting model

Diffusion models, usually logistic models, are used for forecasting the diffusion of new products among a population of potential consumers when the product is at the beginning of its lifecycle (Kunc, 2009). The model has four parameters which can be interpreted in terms of external effects:

(1) (innovativeness) originated from mass media or advertising campaigns (a);
(2) internal effects (persuasion from previous wine tourists) generated by word of mouth (b);
(3) the maximum potential number of wine tourists (m);
(4) the cumulative number of wine tourists in period t ($N(t)$) and
(5) the forgetting rate (at which wine tourists forget their experience) (c).

Since wine tourism is a service and it is not observable by potential adopters, there is a reduction in internal effects over time as wine tourists forget about their wine experience and they do not generate additional word of mouth. The model is:

$$\frac{\mathrm{d}N(\mathrm{t})}{\mathrm{d}t} = \frac{a(m - N(t)) + b}{mN(t)(m - N(t)) - cN(t)}$$

Step 3—Scenarios

Winery managers do not always appreciate the value of having visitors to return but they should encourage visitors to repeat their visit with a certain frequency. One scenario to consider is to include different levels of repeated tourists, those tourists who become enthusiast visitors. The new version of the model with a returning coefficient *(d)* is:

$$\frac{\mathrm{d}N(\mathrm{t})}{\mathrm{d}t} = \frac{a(m - N(t)) + b}{mN(t)(m - N(t)) - cN(t) + \mathrm{d}N(t)}$$

Other scenarios to consider are related to the uncertainties in terms of social changes and economic performance of the country. Using these uncertainties, set of scenarios using Kunc and O'Brien (2017) methodology can be constructed changing the parameters of the model to identify different future paths.

9.6 Conclusion

The literature survey indicates limited development of wine tourism forecasting. To support the development of wine tourism, this paper proposes a framework combining surveys, modelling and scenarios.

The survey results can show the behaviour in terms of adopting wine tourism as a regular consumption experience. When the forgetting rate of the visitors is very high and, consequently, the effect of word of mouth is relatively weak, the wine industry is losing an opportunity to improve brand loyalty by encouraging local wine visitors to frequently visit their wineries.

Forecasting models have limitations imposed by the data employed as well as the possible multiple development paths. First, the survey describes a situation that is not referred to past behaviour, as the survey is based on questions suggesting potential future behaviour. The stated intended behaviour may never materialise in the future without the right incentives. Therefore, the results are mainly exploratory rather than predictive such as in other papers forecasting tourism based on historical time series.

Appendix

Id	Authors	Title	Reference
1	Avanzini E.L., Mac Cawley A.F., Vera J.R., Maturana S	Comparing an expected value with a multistage stochastic optimization approach for the case of wine grape harvesting operations with quality degradation	International Transactions in Operational Research, 2023, volume 30, issue 4, pages 1843–1873
2	Navia-Osorio E.G., Porras-Amores C., Mazarrón F.R., Cañas I	Impact of climate change on sustainable production of sherry wine in nearly-zero energy buildings	Journal of Cleaner Production, 2023, volume 382, 135260
3	Pesavento M.T	The impact of direct to consumer shipping laws on the number and size distribution of U.S. wineries	Journal of Wine Economics, 2022, volume 17, issue 4, pages 270–295
4	Letamendi J., Sevigne-Itoiz E., Mwabonje O	Environmental impact analysis of a Chilean organic wine through a life cycle assessment	Journal of Cleaner Production, 2022, volume 371, 133368
5	Forliano C., Ferraris A., Bivona E., Couturier J	Pouring new wine into old bottles: A dynamic perspective of the interplay among environmental dynamism, capabilities development, and performance	Journal of Business Research, 2022, volume 142, pages 448–463
6	Borrello M., Cembalo L., Vecchio R	Consumers' acceptance of fungus resistant grapes: Future scenarios in sustainable winemaking	Journal of Cleaner Production, 2021, volume 307, 127318

(continued)

(continued)

Id	Authors	Title	Reference
7	Wittwer G., Anderson K	A Model of Global Beverage Markets	Journal of Wine Economics, 2020, volume 15, issue 3, pages 330–354
8	Ferrara C., De Feo G	Comparative life cycle assessment of alternative systems for wine packaging in Italy	Journal of Cleaner Production, 2020, volume 259, 120,888
9	Ramos A., Berzosa J., Clarens F., Marin M., Rouboa A	Environmental and socio-economic assessment of cork waste gasification: Life cycle and cost analysis	Journal of Cleaner Production, 2020, volume 249, 119316
10	Galindro A., Matias J., Cerveira A., Santos C., Marta-Costa A	Prediction of Viticulture Farms Behaviour: An Agent-Based Model Approach	Palgrave Studies of Cross-Disciplinary Business Research, in Association with EuroMed Academy of Business, 2020, pages 155–178
11	Fountain J., Yuan J., Zhai B., Bauman M.J., Williams H.A., Gu Q	Wine tourism development in China	Handbook on Tourism and China, 2020, pages 187–201
12	Ponstein H.J., Ghinoi S., Steiner B	How to increase sustainability in the Finnish wine supply chain? Insights from a country of origin based greenhouse gas emissions analysis	Journal of Cleaner Production, 2019, volume 226, pages 768–780
13	Mandabach K., Chuanbiao W	Gastronomic Festivals and Events: Future scenarios	The Routledge Handbook of Gastronomic Tourism, 2019, pages 481–488
14	Onofri L., Bianchin F., Boatto V., Furlani M., Pecci F., Perali F	What Is the Future of the Wine-Grape Industry in Veneto? Evidence from a Micro-Macro Prediction Model	Journal of Wine Economics, 2018, volume 13, issue 4, pages 384–393

(continued)

(continued)

Id	Authors	Title	Reference
15	Ashenfelter O., Storchmann K	Using hedonic models of solar radiation and weather to assess the economic effect of climate change: The case of Mosel valley vineyards	Handbook of The Economics of Wine, 2018, pages 59–96
16	Sacchelli S., Fabbrizzi S., Bertocci M., Marone E., Menghini S., Bernetti I	A mix-method model for adaptation to climate change in the agricultural sector: A case study for Italian wine farms	Journal of Cleaner Production, 2017, volume 166, pages 891–900
17	Litskas V.D., Irakleous T., Tzortzakis N., Stavrinides M.C	Determining the carbon footprint of indigenous and introduced grape varieties through Life Cycle Assessment using the island of Cyprus as a case study	Journal of Cleaner Production, 2017, volume 156, pages 418–425
18	Corsinovi P., Gaeta D	European wine policies and their consequences on the global wine trade	Economia Agro-Alimentare, 2017, volume 19, issue 1, pages 59–88
19	Varsei M., Polyakovskiy S	Sustainable supply chain network design: A case of the wine industry in Australia	Omega, 2017, pages 236–247
20	Point E., Tyedmers P., Naugler C	Life cycle environmental impacts of wine production and consumption in Nova Scotia, Canada	Journal of Cleaner Production, 2012, volume 27, pages 11–20
21	Teil G	Climate or Technical Change in Wine?: Confronting climatologists' and winegrowers' analyses	The Routledge Handbook of Wine and Culture, 2022, pages 424–433

(continued)

(continued)

Id	Authors	Title	Reference
22	Teruel M., Soldevila-Lafon V., Martin-Bofarull M	A forecast of Cava wine sales applied to vine planting authorizations	British Food Journal, 2022, volume 125, issue 13, pages 1–15
23	Gualberto C.R., Rodrigues L.F., Ferreira K.A	Evaluation of postponement strategy in the production of table wine in Brazil: a discrete event simulation approach	International Journal of Wine Business Research, 2021, volume 33, issue 4, pages 545–560
24	Liu A., Song H	Analysis and Forecasts of the Demand for Imported Wine in China	Cornell Hospitality Quarterly, 2021, volume 62, issue 3, pages 371–385
25	Paroissien E	Forecasting bulk prices of Bordeaux wines using leading indicators	International Journal of Forecasting, 2020, volume 36, issue 2, pages 292–309
26	Hirche M., Haensch J., Lockshin L	Comparing the day temperature and holiday effects on retail sales of alcoholic beverages – a time-series analysis	International Journal of Wine Business Research, 2020, volume 33, issue 3, pages 432–455
27	Voon D., Fogarty J	A Note on Forecasting Alcohol Demand	Journal of Wine Economics, 2019, volume 14, issue 2, pages 208–213
28	Robinson D., Aguirre C., Griffin P	Global climate change influences wine production and tourism in northern New York	Proceedings of the International Conference on Tourism Research, 2019, pages 246–255
29	Fogarty J., Voon D	Alcohol Consumption in the United States: Past, Present, and Future Trends	Journal of Wine Economics, 2018, volume 13, issue 2, pages 121–143
30	De Salvo M., Begalli D., Capitello R., Agnoli L., Tabouratzi E	Determinants of winegrowers' profitability: evidence from an Eastern Europe wine region	EuroMed Journal of Business, 2017, volume 12, issue 3, pages 300–315

(continued)

(continued)

Id	Authors	Title	Reference
31	Jiao L	Macroeconomic determinants of wine prices	International Journal of Wine Business Research, 2017, volume 29, issue 3, pages 234–250
32	Barber N.A., Christopher Taylor D	Experimental approach to assessing actual wine purchase behaviour	International Journal of Wine Business Research, 2013, volume 25, issue 3, pages 203–226
33	Aiello G., Cannizzaro L., La Scalia G., Muriana C	An expert system for vineyard management based upon ubiquitous network technologies	International Journal of Services Operations and Informatics, 2011, volume 6, issue 3, pages 230–247
34	Kunc M.H	Forecasting the development of wine tourism: A case study in Chile	International Journal of Wine Business Research, 2009, volume 21, issue 4, pages 325–338

References

Alonso, A. D., & Liu, Y. (2012). Old wine region, new concept and sustainable development: Winery entrepreneurs' perceived benefits from wine tourism on Spain's Canary Islands. *Journal of Sustainable Tourism, 20*(7), 991–1009.

Connolly, T. M., Boyle, E. A., MacArthur, E., Hainey, T., & Boyle, J. M. (2012). A systematic literature review of empirical evidence on computer games and serious games. *Computers & Education, 59*(2), 661–686.

Hall, M. C., & Mitchell, R. (2000). Wine tourism in the mediterranean. A tool for restructuring and development. *Thunderbird International Business Review, 42*(4), 445–465.

Hall, C. M., & Macionis, N. (1998). Wine tourism in Australia and New Zealand. In R. W. Butler, C. M. Hall, & J. M. Jenkins (Eds.), *Tourism and recreation in rural areas*. Wiley.

Jiao, E., & Chen, J. (2019). Tourism forecasting: A review of methodological developments over the last decade. *Tourism Economics, 25*(3), 469–492.

Kunc, M. H. (2009). Forecasting the development of wine tourism: A case study in Chile. *International Journal of Wine Business Research, 21*(4), 325–338.

Kunc, M. (2018). *Strategic analytics: Integrating management science and strategy*. John Wiley & Sons.

Kunc, M., & O'brien, F. A. (2017). Exploring the development of a methodology for scenario use: Combining scenario and resource mapping approaches. *Technological Forecasting and Social Change, 124*, 150–159.

Mitchell, R., & Hall, C. M. (2006). Wine tourism research: The state of play. *Tourism Review International*, *9*(4), 307–332.

Song, H., & Li, G. (2008). Tourism demand modelling and forecasting-a review of recent research. *Tourism Management, 29*(2), 203–220.

Steinhagen, S., Darroch, J., & Bailey, B. (1998). Forecasting in the wine industry: An exploratory study. *International Journal of Wine Marketing, 10*(1), 13–24.

Torres, J. P., Barrera, J. I., Kunc, M., & Charters, S. (2021). The dynamics of wine tourism adoption in Chile. *Journal of Business Research, 127*, 474–485.

10

Broadening the Idea of Guides and Guiding: A Wine Tourism Roadmap

Terrance G. Weatherbee and Donna Sears

10.1 Introduction

While the art and science of guides and guiding have been the subject of concerted academic focus in general (Weiler & Black, 2015b)—and much of the niche or special interest tourism (Novelli, 2022; Novelli & Humavindu, 2005) literatures—research on guides and the process of guiding in wine tourism has been sparse. Despite four decades of intensive study into wine tourism, the contribution of guides and guiding to wine tourist experiences have largely been overlooked (Robertson et al., 2018; Weatherbee & Sears, 2019, 2021). While there is acknowledgement of guides, it largely takes the form of taken-for-granted mentions in passing where guides are ancillary by-products of the study of some other phenomenon in wine tourism.

T. G. Weatherbee (✉) · D. Sears
FC Manning School of Business, Acadia University, Wolfville, NS, Canada
e-mail: terrance.weatherbee@acadiau.ca

D. Sears
e-mail: donna.sears@acadiau.ca

J. Martínez-Falcó et al. (eds.), *Strategic Management in the Wine Tourism Industry*,
https://doi.org/10.1007/978-3-031-54837-6_10

Markers of this are apparent across our journals. For example, a search of the *International Journal of Wine Business Research* using the Proquest database reveals that of the 741 articles in the database, 221 of the articles contain the word "guide". Of these, the vast majority do have the term somewhere in the main text, but only three have it as a subject term, and only one article has a guide as the subject of research! Similar results were found in other journals covering wine business: the *Journal of Wine Research*, the *International Journal of Wine Research*, *Wine Economics and Policy*, and the *Journal of Wine Economics*.

Our goal in this chapter is to correct this oversight and to set out a conceptual framework for the study of guides and guiding in wine tourism. We believe a separate framework is required to account for the unique characteristics that wine tourism has over other forms of tourism (Hall et al., 2009a). The production and consumption of wine is a complex cultural phenomenon (Inglis & Almila, 2022) which means that there are three factors that any such framework must consider. First, unlike most other forms of tourism, a product is at the core of the wine tourism experience. It follows that wine tourism is simultaneously comprised of both product and experiential characteristics. Thus, the requirement for guides and guiding can be expected to impact the customer journey differentially as the customer focus moves between product and experience. Second, since wine consumption is sociologically context-sensitive (Picard et al., 2018), practices will vary from one culture to the next, so universality in the forms of guides and guide practice cannot be expected. Finally, due to the alcoholic content of wine, regulatory variation between jurisdictions also impacts wine business operations, and consequently guides and guiding practices (Bruwer et al., 2012; Hashimoto & Telfer, 2003; Kotur, 2023; Reynolds & Knowles, 2014).

With these factors in mind, we first reviewed the general tourism, wine business, and wine tourism literatures to identify what artefacts, technologies, and processes are known to guide wine consumption and wine tourism experiences. This permitted us to broadly conceptualize wine guides and guiding within the product-experience and context-sensitive dimensionalities of wine consumption. Together, this yielded a four-quadrant framework within which to identify, situate, and classify wine guides (see Fig. 10.1). As the range of types and forms of guides and guiding is unusual (i.e., product and experience) in comparison to those found in other forms of tourism, we surface the implications of this by relating these quadrants to the customer journey (Lemon & Verhoef, 2016; Yachin, 2018) to identify potentially fruitful areas of enquiry in wine tourism.

In the sections which follow, we commence with a brief history of guides and guiding—which are both historically situated. We then work to expand the notion of what a guide is beyond what has been normatively accepted in the literature. We do so to account for both the unique features of wine tourism and the effects of technology. Technologies are increasingly blurring the agential nature of relationship(s) established between tourists, guides, and destinations during the co-creation of wine tourism experiences (Rachão et al., 2021; Robinson & Sigala, 2019). This conceptual broadening allows us to then map the various guides and guiding practices onto the consumption and experiential nature of the

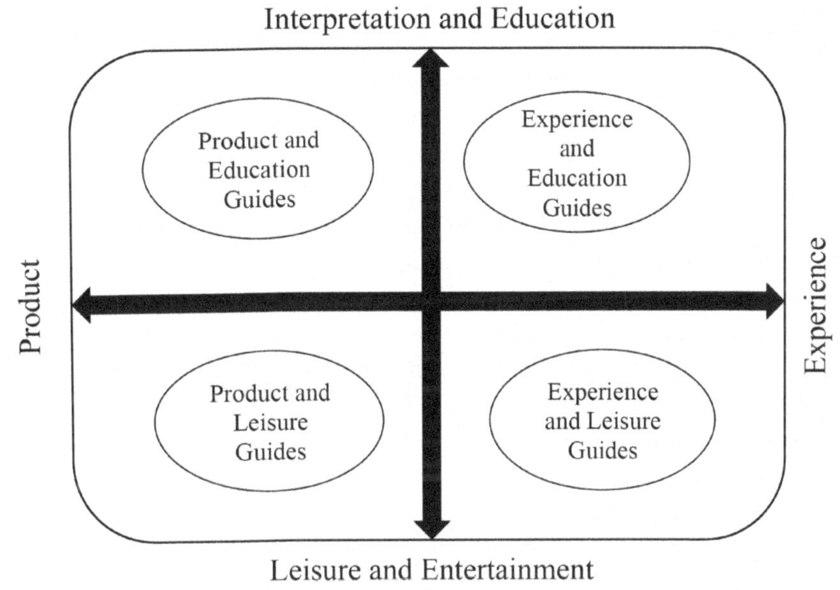

Fig. 10.1 Guide purpose in wine and wine tourism (*Source* Own elaboration)

culturally situated practices of wine tourism. Following this, we discuss the implications that this reconceptualization of guides and guiding has for the wine tourist customer journey.

10.2 A Brief History of Guides and Guiding

Like the growing and making of wine itself, the concept of guides and the practice of guiding are deeply ingrained within human history. Found across all cultures and in all civilizations, guides have played critical roles in helping others to achieve their goals by assisting them in navigating the unfamiliar. Whether travelling and mapping the physical world or exploring and explaining the mental worlds of interpretation and education, for millennia guides have provided assistance, advice, and instruction to those seeking to find their way.

Some of the earliest evidence of the work of guides is found in religious records. In ancient Egypt, for example, priests led pilgrims to sacred sites while freelance guides were to be found surrounding the pyramids waiting to ply their trade on visitors (Hunt, 1984). The historical relationship between guides and religious visitations or pilgrimage has long persisted. In the Middle Ages, the Knights Templar acted as guides for pilgrims travelling to Jerusalem, offering protection and assistance throughout the journey (Forey, 2003). Modern instantiations of guided pilgrimages range from the Hajj to Mecca (Peters, 1996) or the walks of Santiago de Compostela (Slavin, 2003).

Guides and the practice of guiding are also of economic (both trade and tourism) and military importance. In ancient Greece, guides helped travellers cross the treacherous mountain paths of the Greek islands (Della Dora, 2008). Similarly, in China, Sherpas helped traders to navigate the mountain passes of the Himalayas, carrying heavy loads and providing important support along the way (Sherpa & Wengel, 2023). As European nations explored the world beyond their continental boundaries, Arab sailors often guided European merchants throughout ports along the coasts of Africa and the Indian Ocean (Hourani & Carswell, 1995).

For millennia, militaries have employed specialized soldiers, called pathfinders, to mark pathways and guide armies into and across unknown or foreign terrain. Much of the success of Alexander the Great's campaigns was due to his astute use of native guides for pathfinding or as interpreters to gather intelligence needed to understand enemy military capabilities (Engels, 1980a, 1980b). For these reasons the strategic employment of local guides for knowledge of cultural practices and local geography is still relevant in military operations in the twenty-first century (Guzman, 2014).

While guides and guiding have a long history in modern society, for tourism the contemporary guiding profession has its roots in the fifteenth century. Beginning at this time, young men and women from titled families would travel extensively throughout Europe on what became known as the Grand Tour (Towner, 1985). The Grand Tour was the culminating activity of a young aristocrat's education (Brodsky-Porges, 1981). Most were students of fifteen to seventeen years old and were accompanied by tutors on their travels. These tutors would guide, teach, and mentor them on their sojourns; travels often lasting months in duration (Dewar, 2000). When required, local guides would also be employed to assist with sightseeing or teaching about specific locales or regional cultural practices (Sweet, 2012). The demand for guides would grow as the practice of tourism itself grew.

In the mid-nineteenth century, the era of mass tourism commenced with the first train excursions organized by Thomas Cooke (Brendon, 1991). The continued growth of the middle class and the increase in leisure time away from work meant that by the twentieth century mass tourism was becoming a global phenomenon. This growth accelerated dramatically after the Second World War (Butcher, 2020). Throughout this period the practice of guiding began to professionalize, concurrent with the increasing standardization of mass tourism—i.e., packaged tours (Pond, 1993).

The twentieth century not only saw continued growth in mass tourism but also a diversification of tourism products and experience offerings. Niche or special interest tourism, tourism catering to the wants of specific market segments, also become a growth area (Benur & Bramwell, 2015).

While mass tourism guides provided interpretation, assistance, and information to visitors on tours of popular tourist destinations, special interest guides lead eco-tours to expose tourists to the natural environment in parks or conservation areas, or led adventure tours such as mountain climbing or white-water river rafting (Arnould & Price, 1993; Walter, 2013). Still others, specialized in such areas as farm visitations in agritourism (Phillip et al., 2010), or in culinary, beverage and food tourism (Okumus, 2021). Wine tourism grew to become a special interest sub-set of food and beverage tourism (Hall et al., 2004).

The growth and expansion of tourism, in both mass and special interest forms, saw guides and guiding become professionalized and standardized. Over time the training of guides and the standards to which they would practice their craft would increasingly become normalized and then regulated for destinations (e.g., countries and regions) or for the type of tourism site or practice (e.g., museum or adventure) through international or national guilds or associations such as the World Federation of Tourist Guide associations (https://wftga.org).

Finally, one of the most significant developments in the history of tourism, and relatedly of guides and guiding, has been the rise and adoption of the Internet and related technologies (Sigala & Haller, 2023). It has now become easier than ever before to access information and advice on a wide range of places, sites, geographies, and topics (Standing et al., 2014) throughout the customer journey. This has meant that the relationship(s) between guides and tourists, and thus the practices of guiding, has irrevocably changed.

The practices of guiding have historically evolved along two related but distinct lines: that of the *pathfinder* and that of the *mentor* (Cohen, 1985). For pathfinding, guides have been historically fundamental to the movement of people over land and sea; for religious pilgrimage, economic trade, or military operations. Guides have also been integral to mentoring in personal development and education through activities such as exploration, cultural interpretation and mediation, or leisure and entertainment within novel locales. In Wine Tourism, guides perform geographic exploration, education, and interpretation at product and experiential levels in service of the developmental and/or leisure pursuits of wine tourists. So, in the twenty-first century, the historical role of guides and guiding have expanded far beyond its origins (Marson, 2011).

10.3 Reconceptualizing Guides in Wine Tourism

The tourism literature informs us that tour guides have long played play key and critical roles in tourism (Cohen, 1985). They act as leaders, pathfinders, narrators, or storytellers, and as mentors for visiting tourists. Guides provide or facilitate the provision of services to tourists throughout their destination travels. Hence, these agents may be considered as brokers mediating the tourist experience (Weiler & Black, 2015a). In this respect, guiding is multidimensional as guides mediate the physical (travel through landscapes), social (interactions with locals), intellectual (providing knowledge and cultural interpretation), and emotional (facilitating affinity with the landscape or hosts) dimensions of the tourist experience (Weiler & Black, 2015b).

In their intermediatory roles in these tourism experiences (Reisinger & Steiner, 2006; Weiler & Davis, 1993), the behaviour(s) and performance(s) of guides mediate the relationship that tourists have with the offerings of their chosen destination. Empirical research shows the more likeable a guide (e.g., nice, friendly, polite, and pleasant) the more likely that guided tourists will perceive value in tour activities (Tsaur et al., 2021). Moreover, positively perceived professional tour guide performance has been found to be directly related to increased levels of expressed satisfaction (Hosany et al., 2017) and other related outcomes. Positive word of mouth, increased destination loyalty, revisit intention, and greater intent to recommend have all been linked back to guide performance and behaviour (Kim, 2018; Syakier & Hanafiah, 2022). The opposite is also true. Failure to conduct themselves professionally or favourably usually negatively impacts these outcomes.

However, regardless of the wealth of empirical evidence on guides and their importance, the literature on guides is still largely grounded in a fairly narrow view of what a guide is. A view that has not, in general terms, kept up with changes to the global tourism landscape.

This contention is based on the following logic. If we acknowledge that a guided tour is "a form of tourism [with] …direct participation by agents apart from the tourists themselves" (Schmidt, 1979: 441) and if we further accept the evidence that guides mediate the relationship between the tourist and the destination (albeit at multiple levels), then we must ask what other agents may mediate the tourist-experience relationship? This is particularly relevant as tourism experiences are co-created (Weiler & Black, 2015a). As tourism outside of packaged or guided tours—human-guided tourism experiences—are *also* mediated experiences, we must also then ask what this implies for self-guided or technology-guided experiences and what it means to be *guided*.

For wine tourism, we propose the adoption of a broader view of guides and guiding, one that reconceptualizes and expands our notion of what a guide is and includes other tourism-situated agents and guiding practices. For example, the literature on tourism, wine business, and wine tourism identifies various artefacts and processes that act as either pathfinding or mentoring/education guides. It is possible to evaluate these and map them along both the product and experience dimensions of guides and guiding (see Fig. 10.1).

10.4 Dimensions of Wine Experiences

Wine is an agricultural good that is produced world over and sold in a globalized marketplace (Bartlett & Ghoshal, 2000). The immense number of grape varieties and regionalized styles of wine mean that wine is inherently a complex product (Rasmussen & Lockshin, 1999). For consumers, the sheer number, and varieties, of wine available to them is further compounded by the fact that wine is an experience good (Nelson, 1970). That is, the nature and characteristics of wine can only be fully known during the act of its consumption. Thus, selling wine to consumers is a process that has both product and experiential dimensions.

On the one hand, wine is clearly a product. It is a tangible good that is produced, marketed, and sold like any other. In fact, wine is one of the oldest and most widely produced alcoholic products in the world

(Rasmussen, 2014). It is grown, harvested, and processed as a value-added agricultural product. Its quality can be measured and evaluated based on objective criteria, such as taste, aroma, and colour (Reynolds, 2010). While wine may be consumed for its taste, aroma, and the presence of alcohol (and its effects), wine is also associated with socio-cultural acts of socializing, relaxation, or celebration (Charters, 2006). Thus, the attraction of wine as a consumer good is not just about its physical or sensory properties, but also about the context within which it is consumed. Wine can be enjoyed by oneself or along with others, in a formal or in a casual setting, alone or with food as a complement. Thus, the experience of wine consumption is impacted by a wide range of factors beyond the wine itself, such as the occasion, the people present, the food pairing, the atmosphere, or the locale of consumption (Charters, 2006). On the other hand, wine is clearly experiential as well.

The experience of wine consumption is also influenced by the marketing and branding of wine products. Wine labels, bottle designs, and advertising campaigns all contribute to the overall wine consumption experience. The way that a wine is marketed can influence how it is perceived by consumers and how they feel when they consume it (McSweeney et al., 2017). For example, wine that is marketed as a premium product may be perceived as more enjoyable and satisfying as a celebratory drink, than wine that is marketed as an everyday product.

The contextual experience of wine consumption is so important that many wineries and wine shops are focusing on providing unique wine experiences to their customers. For example, wineries may offer wine tastings and tours that allow customers to learn about the wine-making process and to sample various wines in a relaxed and informative setting. Some wineries offer related special events, such as grape stomping or wine and cheese pairing classes, providing customers with a memorable and enjoyable wine experience. As wine businesses around the world make significant efforts to differentiate their products in the marketplace, the variety and complexity of the types and styles of wine available to consumers usually necessitates that education is a part of their offerings (Ali-Knight, 1999). Indeed, interest in learning about wine is one of the prime motivating factors for wine tourists to travel to wine regions and engage in wine experiences (Charters & Ali-Knight, 2000). Simultaneously, for those who are not as interested in learning about the complexities of wine, where it is made or how it is produced, wine may simply be perceived as just another form of leisure with alcohol (Weatherbee & Sears, 2019).

Since wine has both product and experiential characteristics, and given that wine can be enjoyed from both intellectual and sensory perspectives, we can formulate a two-by-two matrix that describes wine consumption on two dimensions: a product-experience dimension and an education-leisure dimension (see Fig. 10.1). Guides and guiding practices may then be mapped onto the quadrants of the matrix.

Product and Education Guides: These guides focus on providing wine consumers with information on wine as well as product, in addition to the cultural and historical context surrounding it. They prioritize cultural interpretation and education over leisure and entertainment. Examples of these types of guides may include books or online resources that provide detailed information about individual wines, the wine-making process, the history of wine production in a region, or the specific practices and cultural significance associated with the consumption of wine.

Product and Leisure Guides: These types of guides prioritize providing wine consumers with an enjoyable experience related to the hedonic consumption of wine. They may include printed guides, apps, or other technological tools that highlight the various types of wine available and with information on how to best enjoy the product. These guides prioritize entertainment and leisure over cultural interpretation and education.

Experience and Education Guides: These guides provide wine tourists with immersive experiences related to wine, such as vineyard tours and winery visits, that emphasize regional or local cultural and wine education. These guides provide wine tourists with a deeper understanding of the wine-making process and the cultural and historical context(s) surrounding its production and consumption in a particular locale.

Experience and Leisure Guides: These guides focus on providing wine consumers with an immersive and enjoyable experience related to wine. While these may also include activities such as wine tastings, vineyard tours, or winery visits, they prioritize entertainment and leisure through celebratory or socialization consumption over education or cultural interpretation.

Reconceptualizing and Mapping Guides

Starting with the product anchor, the branding and labels on wine bottles contribute to customer awareness and are often factors in customer decision-making on purchase and travel (Lockshin & Spawton, 2001).

Those interested in wine often read gourmet magazines, consult wine guides, pursue wine education courses (e.g., sommelier training) to increase their knowledge of wine (Colombini, 2015). Word of Mouth from family, friends, or other wine consumers/tourists also provide information that strongly influences the activities of wine consumption and wine tourism (Hall et al., 2009b; Torres et al., 2021).

Moving towards the experience anchor, wine blogs, both those written by wine entrepreneurs as well as by other wine consumers or wine tourists, represent an additional information resource to guide wine tourists in creating their wine experiences (Colombini, 2015). Regional wine guides may provide both general and experiential information concerning wine tourism areas which are themselves culturally situated (Schreiner & Whittall, 2020).

For self-directed travellers, maps may be downloaded from the Internet or in-car mobile GPS navigation systems (Popp & McCole, 2016), allowing wine tourists to travel freely or in strategically developed wine routes where wayfinding is organized by a Destination Management Organization (DMO) or industry association (Brás et al., 2010; Telfer, 2001). DMOs often provide additional wayfinding guides such as route maps and highway signage (Byrd et al., 2017a). They may also sponsor wine festivals or other events along the route. Festivals may be curated events centred on educating participants in regional wine knowledge or culture, or festivals intended primarily for leisure purposes (Yuan et al., 2005). Leisure tours with celebration and socializing through acts of alcohol consumption, and where educational development is minimized or non-existent, is one area where the cultural and regulatory differences of wine and wine consumption are highlighted (see Fig. 10.2).

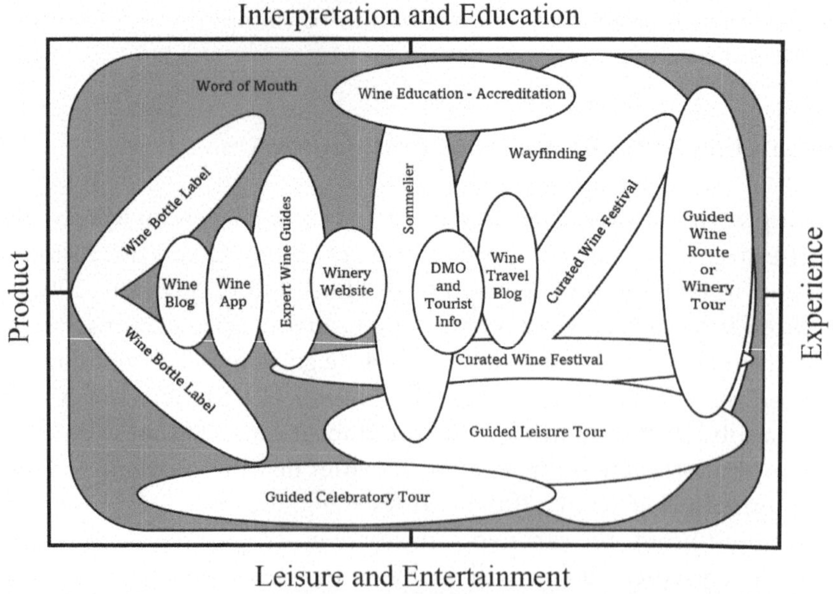

Fig. 10.2 Guides in wine and wine tourism (*Source* Own elaboration)

At the winery level, those that engage with wine tourism activities (e.g., beyond just cellar door sales), there are often guided tastings and tours of the winery facilities and vineyards (Charters & Ali-Knight, 2000; Crespi-Vallbona & Mascarilla-Miró, 2020; Mitchell et al., 2009). Wine consumers and tourists may be given an opportunity to meet the winemaker (Cambourne & Macionis, 2009), see production facilities (Joy, 2018), to observe crushing or fermentation activities (Kotur, 2022) or take part in other complementary activities such as learning about wine and food pairings (Telfer, 2009). Apart from winery tours, many forms of guided wine tours are offered by third-party businesses. These are usually pre-packaged or customized tours that take place within designated regions (Garibaldi et al., 2017) or along organized wine routes (Brás et al., 2010) incorporating more than one winery. These guided or curated tours are designed to introduce wine tourists to a region's landscape and vineyards, wine-growing practices, local architecture, and wine culture, what has been described as a region's winescape (Bruwer &

Alant, 2009; Bruwer & Gross, 2017). Alternatively, rather than a predefined itinerary, some guided tour businesses co-create the tour itinerary with the customer. Finally, sometimes customers create their own wine tourism experience itineraries (Fonseca et al., 2021).

Over the last two decades, technologies of various types are rapidly becoming indispensable tools for both the supply and demand side of the customer journey equation in tourism (Zhang et al., 2018). Websites (Duarte Alonso et al., 2013; Hall et al., 2002; Sears & Weatherbee, 2023) and social media (Sigala & Haller, 2019) are technologies that affect the customer journey, yielding more agency to the customer in terms of acquiring and sharing information needed for decision-making. Mobile devices (Pelet & Lecat, 2014) and apps running on cellphones (Dimitrovski et al., 2019) are increasingly impacting wine tourism throughout the entire customer journey (Sigala et al., 2023) as they may be used to create touchpoints with wineries, for information-seeking, or decision-making on where and what to tour, as well as for the purchase of products and services.

The adoption of more complex technologies, such as virtual, augmented, and mixed reality technologies (Flavián et al., 2019), was accelerated during the COVID-19 pandemic (Gastaldello et al., 2022). Current and near horizon technologies such as artificial intelligence, chatbots, and natural language processing, are in the process of fundamentally impacting and changing the tourism sector overall (Bulchand-Gidumal, 2022). They will also likely significantly impact wine tourism as the use of these technologies will allow tourists to co-create a more personalized consumption experience (Huang et al., 2022; Solakis et al., 2022). Thus we expect that technology will increasingly be deployed in wine tourism (Garibaldi & Dixit, 2022; Saglietto et al., 2016) as Smart Winescapes' potentials become fulfilled (Sears & Weatherbee, 2023).

Guides and the Customer Journey

Following previous research into high involvement (Arnould & Price, 1993) tourism experiences (Lemon & Verhoef, 2016), we use the customer journey to evaluate the relative efficacy and use of product and experiential guides in each stage of the wine tourist journey: pre-trip, destination, and post-trip. Applying a customer journey approach allows the mapping of guide type and use from the micro to macro levels: winery, winescape, and destination scales. It also identifies the various touchpoints wherein guides may mediate tourism, wine tourism, and core wine consumptions, (e.g., travel pathfinding, educational mentoring, cultural interpretation, etc.). Following the customer journey also show where guide use may be freestanding, that is with time and choice of use decided solely by the wine tourist (e.g., use of printed map or GPS for navigation), or co-created where use occurs at touchpoints facilitated by destination product or service-oriented businesses and organizations (e.g., guided tours, DMO publications, winery websites, etc.). Guide use in customer journey stages will differ by the factors of time, space, and touchpoint activity, in relation to the product-experience balance desired.

In the pre-trip phase of the customer journey, wine consumption activities remain product-focused whereas wine tourism activities are largely anticipatory; that is the dominant guide purpose is for information-seeking to aide in destination and activity decision-making, trip planning, and transition-travel related concerns. Information may be sought on the types of wines, or on specific producer/brand using published or online expert guides. Additional information for activity decision-making may involve DMO resources, websites of destination wineries, or other service and activity providers (e.g., accommodations, restaurants, etc.).

Finally, those guides were needed for decisions or arrangements for both transition and destination transportation requirements (e.g., air and ground). During the destination stage, emphasis on guide use shifts as a function of time and space factors and due to the type of activities decided upon and organized during the pre-trip stage. For example, participation in pre-packaged guided tours versus self-guided travel have different requirements in terms of pathfinding (e.g., Google maps, GPS, winery wine route maps) or mentoring (e.g., local wine guides, winery staff, etc.) emphasis. During this stage, the use and utility of guides will normatively shift towards the experiential anchor in comparison to the pre-trip stage, as destination experiences, whether planned or ad

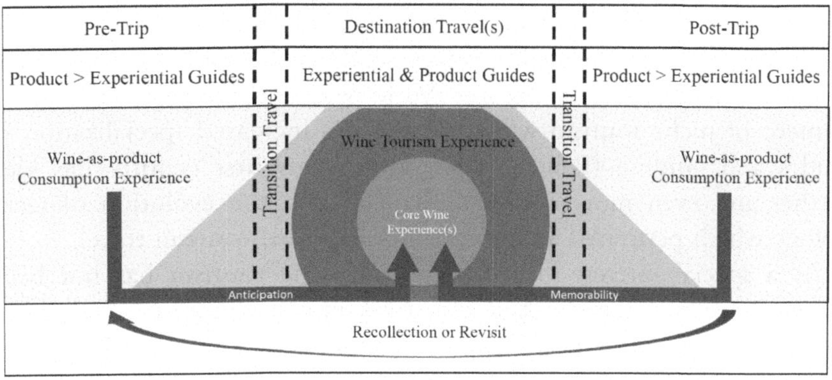

Fig. 10.3 Guides and the customer journey (*Source* Own elaboration)

hoc, become central. Though product-focused guides may still influence travel and activity choice as wine consumption remains at the core of the wine tourism experience (see Fig. 10.3).

In the post-trip phase, guide use will again re-balance, with the use of wine consumption guides again predominating. During this phase, the wine tourist may engage in producing guides themselves, such as contributions to wine blogs, customer reviews on travel sites, etc. Though, these activities may also take place during the destination stage (Sigala & Haller, 2019). It should also be noted that the post-trip stage may also be an information antecedent for subsequent trips (Byrd et al., 2017b). That is, recollection of prior experience may itself act to guide decision-making in the future (Carbone & Haeckel, 1994). This may manifest as the basis for product consumption (e.g., favourite wine choices) or future desired wine tourism experiences (e.g., subsequent destination planning, revisit, or otherwise).

10.5 Conclusion

Guides and the practice of guiding have played an important role in the development of civilization. Throughout the unfolding of human history, guides and guiding practices have continued to evolve to meet the changing needs of individuals, organizations, and whole societies. In the twentieth century, these changes accelerated: first with the advent of mass tourism after the Second World War, and later with the development of niche tourism which necessitates increased specialization in guides and guiding (Pond, 1993). The twenty-first century has seen further and even more dramatic changes with the evolution of technology which permitted an increase in self-guided tourism travel.

As a special interest form of tourism, wine tourism has not been immune to these changes. And wine tourism will continue to be subject to the same trends impacting tourism more generally. These include a consumer desire for experiences over product or service (Schmitt, 1999), the continued enablement of self-guided and co-created tourism experiences (Weiler & Black, 2015a), and the potential for augmenting or outright replacement of guides and guiding practices with technology in both the product and experience dimensions (Tussyadiah, 2020).

Research into guides and guiding in wine tourism lags the understanding of the role and impact of guides as found in the broader tourism literature. Though work on guides and guiding practices in wine tourism remains nascent, the trends affecting tourism overall mean that research in this area will only increase in significance. Both for the academy and for practitioners in wine business, whether in product (wine) or experiential (wine tourism) terms.

References

Ali-Knight, J. (1999). Education in a West Australian wine tourism context. *International Journal of Wine Marketing, 11*(1), 7–18.

Arnould, E., & Price, L. (1993). River magic: Extraordinary experience and the extended service encounter. *Journal of Consumer Research, 20*(1), 24–45.

Bartlett, C., & Ghoshal, S. (2000). Going global: Lessons from late movers. *Reading, 1*(3), 75–84.

Benur, A., & Bramwell, B. (2015). Tourism product development and product diversification in destinations. *Tourism Management, 50*, 213–224.

Brás, J., Costa, C., & Buhalis, D. (2010). Network analysis and wine routes: The case of the Bairrada Wine Route. *The Service Industries Journal, 30*(10), 1621–1641.

Brendon, P. (1991). *Thomas Cook: 150 years of popular tourism.* Secker (Martin) & Warburg Ltd.

Brodsky-Porges, E. (1981). The grand tour travel as an educational device 1600–1800. *Annals of Tourism Research, 8*(2), 171–186.

Bruwer, J., & Alant, K. (2009). The hedonic nature of wine tourism consumption: An experiential view. *International Journal of Wine Business Research, 21*(3), 235–257. https://doi.org/10.1108/17511060910985962

Bruwer, J., & Gross, M. (2017). A multilayered macro approach to conceptualizing the winescape construct for wine tourism. *Tourism Analysis, 22*(4), 497–509.

Bruwer, J., Lesschaeve, I., & Campbell, B. (2012). Consumption dynamics and demographics of Canadian wine consumers: Retailing insights from the tasting room channel. *Journal of Retailing and Consumer Services, 19*(1), 45–58.

Bulchand-Gidumal, J. (2022). Impact of artificial intelligence in travel, tourism, and hospitality. In Z. Xiang, M. Fuchs, U. Gretzel, & W. Hopken (Eds.), *Handbook of e-tourism* (pp. 1943–1962). Springer.

Butcher, J. (2020). Constructing mass tourism. *International Journal of Cultural Studies, 23*(6), 898–915.

Byrd, E., Bhadury, J., & Troy, S. (2017a). Wine tourism signage programs in the USA. *International Journal of Wine Business Research, 29*(4), 457–483. https://doi.org/10.1108/IJWBR-04-2017-0024

Byrd, E., Canziani, B., Boles, J., Williamson, N., & Sonmez, S. (2017b). Wine tourist valuation of information sources: the role of prior travel. *International Journal of Wine Business Research, 29*(4), 416–433. https://doi.org/10.1108/IJWBR-04-2017-0020

Cambourne, B., & Macionis, N. (2009). Meeting the wine-maker: wine tourism product development in an emerging wine region. In *Wine tourism around the world* (pp. 81–101): Routledge.

Carbone, L., & Haeckel, S. (1994). Engineering customer experiences. *Marketing Management, 3*(3), 8–19.

Charters, S. (2006). *Wine and society: The social and cultural context of a drink.* Butterworth-Heinemann.

Charters, S., & Ali-Knight, J. (2000). Wine tourism—A thirst for knowledge? *International Journal of Wine Marketing, 12*(3), 70–80. https://doi.org/10.1108/eb008715

Cohen, E. (1985). The tourist guide: The origins, structure and dynamics of a role. *Annals of Tourism Research, 12*(1), 5–29.

Colombini, D. (2015). Wine tourism in Italy. *International Journal of Wine Research, 7*, 29–35.

Crespi-Vallbona, M., & Mascarilla-Miró, O. (2020). Wine lovers: Their interests in tourist experiences. *International Journal of Culture, Tourism and Hospitality Research, 14*, 239–258.

Della Dora, V. (2008). Mountains and memory: Embodied visions of ancient peaks in the nineteenth-century Aegean. *Transactions of the Institute of British Geographers, 33*(2), 217–232.

Dewar, K. (2000). An incomplete history of interpretation from the Big Bang. *International Journal of Heritage Studies, 6*(2), 175–180.

Dimitrovski, D., Joukes, V., Rachão, S., & Tibério, M. (2019). Wine tourism apps as wine destination branding instruments: Content and functionality analysis. *Journal of Hospitality and Tourism Technology, 10*(2), 136–152.

Duarte Alonso, A., Bressan, A., O'Shea, M., & Krajsic, V. (2013). Website and social media usage: Implications for the further development of wine tourism, hospitality, and the wine sector. *Tourism Planning & Development, 10*(3), 229–248.

Engels, D. (1980a). Alexander's intelligence system. *The Classical Quarterly, 30*(2), 327–340.

Engels, D. (1980b). *Alexander the Great and the logistics of the Macedonian army*. University of California Press.

Flavián, C., Ibáñez-Sánchez, S., & Orús, C. (2019). The impact of virtual, augmented and mixed reality technologies on the customer experience. *Journal of Business Research, 100*, 547–560.

Fonseca, M., Zuanazzi, L., Castilhos, R., & Trez, G. (2021). Value creation of tourist experiences in spectacular and spontaneous designs: A comparative analysis of a wine tour. *Journal of Hospitality and Tourism Insights, 4*(3), 335–350.

Forey, A. (2003). The charitable activities of the templars. *Viator, 34*, 109–141.

Garibaldi, R., & Dixit, S. (2022). The role of technology in wine tourism. In *Routledge handbook of wine tourism* (pp. 557–566). Routledge International.

Garibaldi, R., Stone, M., Wolf, E., & Pozzi, A. (2017). Wine travel in the United States: A profile of wine travellers and wine tours. *Tourism Management Perspectives, 23*, 53–57.

Gastaldello, G., Giampietri, E., Zaghini, E., & Rossetto, L. (2022). Virtual wine experiences: Is Covid extending the boundaries of wine tourism? *Wine Economics and Policy*.

Guzman, L. (2014). *"First in... last out": History of the US army pathfinder (1942–2011).* Retrieved from Fort Leavenworth, Texas.
Hall, C., Johnson, G., Cambourne, B., Macionis, N., Mitchell, R., & Sharples, L. (2009a). Wine tourism: An introduction. In *Wine tourism around the world* (pp. 1–23). Routledge.
Hall, C., Johnson, G., & Mitchell, R. (2002). Wine tourism and regional development. In C. Hall, L. Sharples, B. Cambourne, & N. Macionis (Eds.), *Wine tourism around the world: Development, management and markets* (pp. 196–225). Butterworth-Heinemann.
Hall, C., Longo, A., Mitchell, R., & Johnson, G. (2009b). Wine tourism in new Zealand. In *Wine tourism around the world* (pp. 150–174): Routledge.
Hall, C., Sharples, L., Mitchell, R., Macionis, N., & Cambourne, B. (2004). *Food tourism around the world*: Routledge.
Hashimoto, A., & Telfer, D. (2003). Positioning an emerging wine route in the Niagara region: Understanding the wine tourism market and its implications for marketing. *Journal of Travel & Tourism Marketing, 14*(3–4), 61–76.
Hosany, S., Prayag, G., Van Der Veen, R., Huang, S., & Deesilatham, S. (2017). Mediating effects of place attachment and satisfaction on the relationship between tourists' emotions and intention to recommend. *Journal of Travel Research, 56*(8), 1079–1093.
Hourani, G., & Carswell, J. (1995). *Arab seafaring in the Indian Ocean in ancient and early medieval times.* Princeton University Press.
Huang, A., Chao, Y., de la Mora Velasco, E., Bilgihan, A., & Wei, W. (2022). When artificial intelligence meets the hospitality and tourism industry: An assessment framework to inform theory and management. *Journal of Hospitality and Tourism Insights, 5*(5), 1080–1100.
Hunt, E. (1984). Travel, tourism and piety in the Roman Empire: A context for the beginnings of Christian pilgrimage. *Echos du monde classique: Classical Views, 28*(3), 391–417.
Inglis, D., & Almila, A. (2022). Introduction to the sociology of wine—The sociological decanter: Pouring sociology into wine, and vice versa. *Journal of Cultural Analysis and Social Change, 6*(6), 1–21.
Joy, A., Belk, R., Charters, S., Wang, J. n. F., & Peña, C. (2018). Performance theory and consumer engagement: Wine-tourism experiences in South Africa and India. In *Consumer culture theory*: Emerald Publishing Limited.
Kim, J. (2018). The impact of memorable tourism experiences on loyalty behaviors: The mediating effects of destination image and satisfaction. *Journal of Travel Research, 57*(7), 856–870.

Kotur, A. (2022). Exploring the wellness dimensions of wine tourism experiences: A netnographic approach. *International Journal of Wine Business Research, 35*. https://doi.org/10.1108/IJWBR-07-2021-0040

Kotur, A. (2023). Regulatory and ethical influences and predicaments of wine tourism development. In S. Dixit (Ed.), *Routledge handbook of wine tourism*. New York.

Lemon, K., & Verhoef, P. (2016). Understanding customer experience throughout the customer journey. *Journal of Marketing, 80*(6), 69–96.

Lockshin, L., & Spawton, T. (2001). Using involvement and brand equity to develop a wine tourism strategy. *International Journal of Wine Marketing, 13*(1), 72–81. https://doi.org/10.1108/eb043371

Marson, D. (2011). From mass tourism to niche tourism. In P. Robinson, S. Heitmann, & P. Dieke (Eds.), *Research themes for tourism* (pp. 1–15). CABI.

McSweeney, M., Sears, D., Weatherbee, T., & Jantzi, H. (2017). *The preference for pleasure: Exploring the multidimensional nature of the hedonic sensory experience*. Paper presented at the Pangborn Sensory Science Symposium, Rhode Island.

Mitchell, R., Hall, C., & McIntosh, A. (2009). Wine tourism and consumer behaviour. In *Wine tourism around the world* (pp. 115–135): Routledge.

Nelson, P. (1970). Information and consumer behavior. *Journal of Political Economy, 78*(2), 311–329.

Novelli, M. (2022). Niche tourism. In D. Buhalis (Ed.), *Encyclopedia of tourism management and marketing* (pp. 344–347). Edward Elgar Publishing.

Novelli, M., & Humavindu, M. (2005). *Niche tourism: Contemporary issues, trends and cases* (Vol. 171). Elsevier Ltd.

Okumus, B. (2021). Food tourism research: A perspective article. *Tourism Review, 76*(1), 38–42.

Pelet, J. E., & Lecat, B. (2014). Smartphones and wine consumers: A study of Gen-Y. *International Journal of Wine Business Research, 26*(3), 188–207.

Peters, F. (1996). *The Hajj: The Muslim pilgrimage to Mecca and the holy places*. Princeton University Press.

Phillip, S., Hunter, C., & Blackstock, K. (2010). A typology for defining agritourism. *Tourism Management, 31*(6), 754–758.

Picard, D., Moreira, C., & Loloum, T. (2018). Wine magic: Consumer culture, tourism, and terroir. *Journal of Anthropological Research, 74*(4), 526–540.

Pond, K. (1993). *The Professional guide: The dynamics of tour guiding*. Van Nostrand Reinhold.

Popp, L., & McCole, D. (2016). Understanding tourists' itineraries in emerging rural tourism regions: The application of paper-based itinerary mapping methodology to a wine tourism region in Michigan. *Current Issues in Tourism, 19*(10), 988–1004.

Rachão, S., de Jesus Breda, Z., de Oliveira Fernandes, C., & Joukes, V. (2021). Drivers of experience co-creation in food-and-wine tourism: An exploratory quantitative analysis. *Tourism Management Perspectives, 37*, 100783.

Rasmussen, M., & Lockshin, L. (1999). Wine choice behaviour: The effect of regional branding. *International Journal of Wine Marketing, 11*(1), 36–46. https://doi.org/10.1108/eb008689

Rasmussen, S. (2014). *The quest for Aqua Vitae: The history and chemistry of alcohol from antiquity to the Middle Ages*. Springer Science & Business.

Reisinger, Y., & Steiner, C. (2006). Reconceptualising interpretation: The role of tour guides in authentic tourism. *Current Issues in Tourism, 9*(6), 481–498.

Reynolds, A. (2010). *Managing wine quality: Viticulture and wine quality*. Elsevier.

Reynolds, D., & Knowles, C. (2014). Do US state laws affect per capita wine purchases? *Journal of Foodservice Business Research, 17*(1), 19–27.

Robertson, J., Sears, D., & Weatherbee, T. (2018). *The Wolfville magic winery bus: A grounded theory approach to understanding the consumer view of winescape*. Paper presented at the Atlantic School of Business Conference, Moncton, New Brunswick, Canada.

Robinson, R., & Sigala, M. (2019). Epilogue: An ecosystems framework for studying wine tourism: Actors, co-creation processes, experiences and outcomes. *Wine Tourism Destination Management and Marketing: Theory and Cases*, 603–612.

Saglietto, L., Fulconis, F., Bédé, D., De Almeida Goes, J., & Forradellas, R. (2016). *Wine industry supply chain: International comparative study using social networks analysis*. Paper presented at the Supply Chain Forum: An International Journal.

Schmidt, C. (1979). The guided tour: Insulated adventure. *Urban Life, 7*(4), 441–467.

Schmitt, B. (1999). *Experiential marketing*. The Free Press.

Schreiner, J., & Whittall, L. (2020). *The Okanagan wine tour guide*. TouchWood Editions.

Sears, D., & Weatherbee, T. (2023). The future of the wine tourism experience: The potential of smart (er) winescapes? In *Technology advances and innovation in wine tourism: New managerial approaches and cases* (pp. 135–150). Springer.

Sherpa, S., & Wengel, Y. (2023). *The Sherpas and their original identity.* Cambridge Scholars Publishing.
Sigala, M., Dimitrovski, D., & Joukes, V. (2023). Design factors of mobile app in wine tourism: Creating customer value through the whole wine tourism journey. In *Technology advances and innovation in wine tourism: New managerial approaches and cases* (pp. 47–58). Springer.
Sigala, M., & Haller, C. (2019). The impact of social media on the behavior of wine tourists: A typology of power sources. In *Management and marketing of wine tourism business* (pp. 139–154). Springer.
Sigala, M., & Haller, C. (Eds.). (2023). *Technology advances and innovation in wine tourism: New managerial approaches and cases.* Springer Nature.
Slavin, S. (2003). Walking as spiritual practice: The pilgrimage to Santiago de Compostela. *Body & Society, 9*(3), 1–18.
Solakis, K., Katsoni, V., Mahmoud, A., & Grigoriou, N. (2022). Factors affecting value co-creation through artificial intelligence in tourism: A general literature review. *Journal of Tourism Futures.* https://doi.org/10.1108/JTF-06-2021-0157
Standing, C., Tang-Taye, J., & Boyer, M. (2014). The impact of the Internet in travel and tourism: A research review 2001–2010. *Journal of Travel & Tourism Marketing, 31*(1), 82–113.
Sweet, R. (2012). *Cities and the grand tour: The British in Italy, c. 1690–1820.* Cambridge University Press.
Syakier, W., & Hanafiah, M. (2022). Tour guide performances, tourist satisfaction and behavioural intentions: A study on tours in Kuala Lumpur city centre. *Journal of Quality Assurance in Hospitality & Tourism, 23*(3), 597–614.
Telfer, D. (2001). From a wine tourism village to a regional wine route: An investigation of the competitive advantage of embedded clusters in Niagara, Canada. *Tourism Recreation Research, 26*(2), 23–33.
Telfer, D. (2009). The northeast wine route: Wine tourism in Ontario, Canada and New York State. In *Wine tourism around the world* (pp. 253–271): Routledge.
Torres, J., Barrera, J., Kunc, M., & Charters, S. (2021). The dynamics of wine tourism adoption in Chile. *Journal of Business Research, 127*, 474–485.
Towner, J. (1985). The grand tour: A key phase in the history of tourism. *Annals of Tourism Research, 12*(3), 297–333.
Tsaur, S., Yang, T., & Tsai, C. (2021). Tour leader likeability and tourist citizenship behaviours: Mediating effect of perceived value. *Current Issues in Tourism, 24*(18), 2628–2642.

Tussyadiah, I. (2020). A review of research into automation in tourism: Launching the Annals of Tourism Research Curated Collection on Artificial Intelligence and Robotics in Tourism. *Annals of Tourism Research, 81*, 1–13.

Walter, P. (2013). Theorising visitor learning in ecotourism. *Journal of Ecotourism, 12*(1), 15–32.

Weatherbee, T., & Sears, D. (2019). *Tasted or wasted? Drunkenness and other wine-related things we may have missed!* Paper presented at the Academy of Wine Business Research, Stellenbosch, South Africa.

Weatherbee, T., & Sears, D. (2021). *Making memorable experiences: Tour guides in wine tourism*. Paper presented at the Academy of Wine Business Research, Virtual Conference.

Weiler, B., & Black, R. (2015a). The changing face of the tour guide: One-way communicator to choreographer to co-creator of the tourist experience. *Tourism Recreation Research, 40*(3), 364–378.

Weiler, B., & Black, R. (2015b). *Tour guiding research: Insights, issues and implications*. Channel View Publications.

Weiler, B., & Davis, D. (1993). An exploratory investigation into the roles of the nature-based tour leader. *Tourism Management, 14*(2), 91–98.

Yachin, J. (2018). The 'customer journey': Learning from customers in tourism experience encounters. *Tourism Management Perspectives, 28*, 201–210.

Yuan, J., Cai, L., Morrison, A., & Linton, S. (2005). An analysis of wine festival attendees' motivations: A synergy of wine, travel and special events? *Journal of Vacation Marketing, 11*(1), 41–58.

Zhang, H., Gordon, S., Buhalis, D., & Ding, X. (2018). Experience value cocreation on destination online platforms. *Journal of Travel Research, 57*(8), 1093–1107.

Part V

Epilogue

11

Analysing the Experience of Wine Tourists: An Approach from the Wine Routes

Ricardo Sellers-Rubio, Juan Luis Nicolau, and Seunghun Shin

11.1 Introduction

The wine industry has made great efforts to foster wine tourism as a means of improving brand image, to promote wine culture and, in particular, to increase the income of wineries. Moreover, from an economic point of view, wine-producing regions have concluded that

R. Sellers-Rubio
Departamento de Marketing, University of Alicante, San Vicente del Raspeig, España
e-mail: jnicolau@vt.edu

J. L. Nicolau
Howard Feiertag Department of Hospitality and Tourism Management, Pamplin College of Business, Virginia Tech, Blacksburg, VA, USA
e-mail: ricardo.sellers@ua.es

S. Shin (✉)
School of Hotel and Tourism Management, The Hong Kong Polytechnic University, Kowloon, Hong Kong
e-mail: seung-hun.shin@polyu.edu.hk

J. Martínez-Falcó et al. (eds.), *Strategic Management in the Wine Tourism Industry*,
https://doi.org/10.1007/978-3-031-54837-6_11

converting themselves into new tourist destinations can aid their overall development (Marzo-Navarro & Pedraja-Iglesias, 2010). Thus, beyond the benefits that wine tourism brings to individual wineries, it can provide advantages to all actors in wine-growing areas, such as restaurants, accommodation companies, and other complementary offerings.

Consequently, public/private bodies, such as Wine Routes of Spain (Rutas del Vino de España https://wineroutesofspain.com) and the Spanish Association of Wine Cities (ACEVIN—https://acevin.es), have taken, in collaboration with other stakeholders in the industry, various actions in recent years to encourage and promote high standard, wine-focused, tourism.

From the perspective of destination marketing, wine routes operate as DMOs, which promote the images of specific tourist destinations. Tahar et al. (2021) highlighted the central role of wine-focused DMOs in the coordination of coherent, collective wine tourism strategies, and in the creation of destination brands that encompass the whole wine tourism offer (e.g. wineries, accommodation, restaurants) of a given region. Wine tourism destinations are marketed under collective brands created and managed by different wine routes/DMOs (e.g. the Rioja Wine Route), which play an intermediary role between the services provided by individual wineries and the market. DMOs aim to develop uniform and recognised collective destination brands to act as catalysts for the promotion of quality wine tourism for the benefit of individual wineries.

To understand the interaction between destination brands and wine tourists, it is essential to understand the behaviours of wine tourists who participate in this type of tourism. In general, wine tourism experiences are formed by the specific activities undertaken at the wineries (Angelini & Gilli, 2021; Charters & Fountain, 2006) and at the destinations themselves (Byrd et al., 2016; Cardinale et al., 2016). Leri and Theodoridis (2021) examined the wine tourism experience holistically, as a multidimensional experience composed of the specific activities that the wine tourist undertakes in the winery (e.g. interaction with staff, the sensory experience of the wines tasted and visiting the winery facilities), and an environmental component (winescape), which is related to where the activities take place, that is, the destination and the territory where the winery is located. However, despite the potential advantages of wine routes, many wineries still adopt wine tourism strategies independent of the routes. The key reasons wineries adopt independent strategies are their belief that the routes are "a tool that does not contribute much" and the "lack of professionalism of the managers" of the routes (WRA, 2021).

The goal of this study is to increase the knowledge of wine tourism by analysing wine tourists' experiences to identify: the effects that membership of a wine route have on individual wineries; the effects that visit type have on individual wineries; any differences in tourists' evaluations of wineries belonging to the different routes.

11.2 Theoretical Framework

Much research into wine tourism has taken place in recent years (Gómez et al., 2019); this can be grouped into three main research streams (Vo Thanh & Kirova, 2018).

First, some authors have focused on the stakeholders involved in wine tourism and the strategies followed by the wineries themselves (Dawson et al., 2011). For example, Charters and Ali-Knight (2002) and Veres et al. (2008) highlighted the importance of establishing tasting rooms in wineries to enhance the educational component of tourists' visits.

A second stream of research has focused on destination strategies, examining the effects of these strategies on destination image, examining the initiatives taken by destinations to strengthen their promotion of wine tourism, and carried out cost–benefit analyses (Hojman & Hunter-Jones, 2012). Some of these studies, focused on rural areas, analysed: the socio-economic impact of wine tourism (Alonso & Liu, 2012; Carlsen, 2004); wine routes (Bruwer, 2003; López-Guzmán et al., 2014; Telfer, 2001); the determining factors for the success of wine-producing tourist destinations (Getz & Brown, 2006; Gross & Brown, 2008); and the impact of local wine festivals.

The third stream has focused on the tourist's perspective and has tried to characterise the profile of the wine tourist (Charters & Ali-Knight, 2002; Getz & Brown, 2006) and to identify different wine tourist segments (Alonso et al., 2007; Dodd, 1995; Mitchell & Hall, 2006; Williams & Dossa, 2003). Other topics analysed were consumers' motivations for visiting wine destinations (Carlsen & Boksberger, 2015), visitor satisfaction (Mitchell & Hall, 2006) and the dimensions that help to predict tourists' future behaviours (Sparks, 2007), such as their spending in wineries (Sellers & Nicolau, 2021).

The present study integrates the last two streams of research, that is, by focusing on the effects that the different wine routes have on the wine tourist's experience, working on the assumption that different routes will affect tourists in distinct ways, and to various extents.

While the wineries themselves are, clearly, the essential element in wine tourism, with the visits they arrange to their facilities, another key to the future development, and increased success, of this form of tourism is the general attractiveness of the overall destination, including the natural environment, the climate, the complementary offers of accommodation and gastronomy and the possibility of participating in local cultural activities (Marzo-Navarro & Pedraja-Iglesias, 2010).

Indeed, previous wine tourism-focused research has highlighted the importance of destination image in shaping the tourist experience. Some authors have recognised the added value of tourism destination brands and, therefore, the importance of building successful collective, destination brands, and in identifying the factors that determine destination brand value (Boo et al., 2009; Gómez et al., 2015b; Konecnik & Gartner, 2007). Thus, while the success of individual wineries and their wines is partly based on their brand images (Getz & Brown, 2006), successful wine tourism development is also based on destination brand image, which undoubtedly encourages wine tourists to visit particular wine regions (Gómez et al., 2015b).

In general, the literature argues that to develop a wine route, the individual, constituent wineries must create wine destination organisations, that is, wine-focused DMOs (Hojman & Hunter-Jones, 2012). Gatti and Incerti (1997: 219) defined a wine route as "a sign-posted itinerary, through a limited area (region, province, appellation area) whose aim is the discovery of the wine(s) product(s) in the region and the activities which are associated with it".

In this sense, it should be remembered that wine tourism is the result of the symbiosis between tourism and the wine sector, in which the importance of the producing regions is represented at brand level by "designations (or Appellations) of origin" (DOs) (Hall, 2002; Hall & Macionis, 1998). In fact, the wine routes marketed by wine tourism destinations are usually associated with a wine designation of origin, which in itself is a brand (collective brand), and a key asset for the wineries located in the area covered by the designation. Collective brands, both the wine route itself and the designation of origin, help identify wine tourism destinations, differentiate them from their competitors and develop and strengthen the consumer-wine tourism destination emotional bond (Gómez et al., 2015a). Indeed, there is evidence that the brand image of wine denominations of origin (Gómez et al., 2015b), which have been shown to be useful tools for attracting visitors to wineries, have an impact on tourists (Caldwell & Freire, 2004). The competitive positioning of wine tourism regions is a key strategic issue for destinations (Williams & Kelly, 2001), and many regions deploy strategies to attract wine tourists to increase the financial returns of local businesses (Getz & Brown, 2006).

Bruwer (2003), along the same lines, argued that the branding of a wine region's attractiveness to the market is based on the "difference of place" it projects. Hall (2002) showed that visits to wine regions are motivated often by the region's attributes, their "winescape", which has been described as being composed of three main elements: the presence of vineyards, winemaking activities and the cellars where wine is produced and stored (Telfer, 2001). Similarly, Thomas et al. (2018) highlighted the importance of destination image, based on this same winescape concept. Hall and Mitchell (2002) used the term "tourism terroir" to describe the combination of physical, cultural and natural environments that gives each region its distinctive appeal as a wine tourism destination. This tourism terroir concept broadens the notion of winescape to encompass the feeling of "region", which encompasses the area's emotional appeal, and is the result of the combination of its physical and cultural components (Sparks, 2007). Famularo et al. (2010) emphasised the significance for wines of their region of origin. The combination of a region's environmental, climatic and territorial factors, which are often associated with a single region, define the unique sensory and tactile characteristics of its wines. Moreover, a wine's region of origin plays a decisive role in the consumer's decision-making processes. Therefore, although only limited literature has focused on the attractiveness of wine regions, Charters and Ali-Knight (2002) argued that wine tourists' expectations are likely to vary from one region to another, given the differences between regions, so the importance of a destination's attributes in the choices tourists make, in terms of which wine tourism activities to undertake, cannot be ignored (Getz, 2000; Getz & Brown, 2006), as they might influence the wine tourist's experience. Therefore, this chapter focuses on analysing the wine tourist's experience from the perspective of the wine route to which the wineries visited belong.

11.3 Methods

To achieve the goal of this chapter we analysed the opinions that wine tourists who had visited Spanish wineries posted on TripAdvisor. The methodology used is as follows.

First, we downloaded information about Spanish wineries published on TripAdvisor between 2011 and 2021. We identified 881 wineries, but as 84 had received no comments or ratings from wine tourists, the final sample used was 797 wineries. We collected information about their locations, the total number of reviews they had received, and the aggregate scores (rating) given by the TripAdvisor reviewers. We downloaded 47,344 individual reviews of wineries from wine tourists, of which 35,648 were written in Spanish, and 11,696 in English. Based on previous literature in this field, we excluded very short reviews (i.e. less than ten words), because they generally do not contain meaningful information (Gao et al., 2018). For each of the reviews, the following information was identified: the quantitative numerical score (rating) that the reviewer gave to the experience, date of the visit, qualitative assessment of the visit (written commentary in text form) and type of visit (solo, with friends, as a couple, as a family or for business/work reasons).

Second, a sentiment-based analysis of the qualitative reviews was carried out using the AFINN lexicon (available in Spanish and English), a database of 2477 words expressing degrees of positive and negative sentiment. The lexicon, developed by Finn Arup Nielsen, has previously been used in the wine tourism field (Barbierato et al., 2021). It rates words in the range −5 (most negative sentiment) to +5 (most positive sentiment). The Spanish adaptation of the lexicon was refined by the authors of the present study. In the present work, the sentiment values of the reviews were obtained by summing the polarity of the words used. This additive procedure provides a more fine-grained analysis of the ratings than does the simple classification as positive, negative or neutral offered by other lexicons. Prior to applying the lexicon, the text was divided into sentences, and stop words were removed. In addition, in examining individual words, a proximity cluster of four preceding words, and one following word, was analysed to assess the effect of amplifiers (e.g. "the winery is very nice" has a higher sentiment than "the winery is nice") and/or semantic twists that could alter the meanings of the words (e.g. "the winery is not nice at all" has a negative sentiment).

Third, we obtained the listings of the wineries belonging to the 32 official Spanish wine routes in 2021 from the websites of each route and linked the routes to the wineries in the sample. Of the 797 wineries included in the sample, 341 are associated with a route. Of the 456 not associated with routes, 426 are located in regions where an official wine route exists.

In the last stage, the differences in the evaluations given by tourists to the wineries were analysed based on visit type, the language of the review and the specific wine route to which the visited wineries belonged.

The analysis, processing and estimation of the results were carried out with R programming language.

11.4 Results

First, we analysed the temporal pattern of the opinions posted by the wine tourists. Table 11.1 shows an increasing trend in the number of reviews posted from 2011, reaching its maximum in 2017. A significant decrease in the number of reviews posted took place in the years 2020 and 2021, undoubtedly as a consequence of the COVID-19 pandemic,

Table 11.1 Evolution of the number of wine tourism reviews published on TripAdvisor

Year	Total	Spanish	English
Undated	901	879	22
2011	217	133	84
2012	815	487	328
2013	1241	785	456
2014	2131	1460	671
2015	4875	3493	1382
2016	7617	5776	1841
2017	8452	6408	2044
2018	7638	5601	2037
2019	7405	5463	1942
2020	3115	2691	424
2021	2937	2472	465
Total	47,344	35,648	11,696

Source The authors

lockdown and the associated restrictions on personal mobility, which had a devastating impact on tourism (Fotiadis et al., 2021).

Second, we estimated the quantitative and sentiment-based ratings received by the wineries over the period (see Table 11.2). The average quantitative rating, on scales of 1 to 5, is 4.631 (SD = 0.429), which shows that the wine tourists were highly satisfied with their visits to the wineries. Moreover, the average ratings show a steady improvement over time, from 4.333 in 2011 to 4.718 in 2021. To identify whether differences in evaluation arose based on the nationality of the reviewers, we carried out separate analyses of the Spanish and English reviews. The average score of the Spanish reviews was 4.626 (SD = 0.434), slightly lower than the average score of the English reviews (4.705; SD = 0.514). This difference is statistically significant ($t = -3.26$; $p < 0.01$). The sentiment-based analysis of the qualitative reviews showed that the average rating was 5.883 (SD = 3.371) for the Spanish reviews, and 10.917 (SD = 5.139) for the English reviews. The sentiment-based ratings of the English reviews were, once again, higher than the ratings of the Spanish reviews, and the difference is statistically significant ($t = -20.9$; $p < 0.01$). Overall, these results suggest that international tourists were more satisfied with their visits to the wineries in our sample than were national tourists.

Next, we analysed the ratings given to the wineries based on whether or not they belonged to a wine route. The average of the quantitative ratings given to wineries belonging to wine routes (4.642; SD = 0.377) was slightly higher than that given to wineries not associated with wine routes (4.622; SD = 0.464), although the difference is not statistically significant ($t = 0.654$; $p = 0.514$). It should be remembered that one of the functions of the routes, together with promotion and marketing, is to promote to wineries the quality and product criteria established by ACEVIN and the Secretary of State for Tourism in the Spanish Wine Routes Tourism Product Manual. However, whether a winery belongs to a wine route, or not, does not seem to make tourists more, or less, satisfied with their visits.

Third, the potential influence of visit type on the ratings was analysed. As can be seen in Table 11.3, significant differences exist between the quantitative ratings based on visit type (solo, with friends, with family, as a couple or for work/business reasons). These differences are statistically significant ($F = 4.934$; $p < 0.01$). Significant differences were also observed between reviews posted in Spanish ($F = 4.140$; $p < 0.01$) and in English ($F = 3.636$; $p < 0.01$). As can be seen, most wine tourists posting reviews on TripAdvisor travelled as a couple, and couples awarded higher ratings than the other Spanish visitors. However, the best ratings for the English language reviews were given by those visiting for work/business purposes. The results are similar for the sentiment-based ratings.

We also analysed wine tourists' opinions by wine route. We used the average of the quantitative ratings of the wineries belonging to the routes as proxies for tourist satisfaction with each route. To achieve a representative average of the route ratings, at least 3 of the wineries of a route had to be in the database. From the initial 32 routes, this meant eliminating 5 (WR3, WR9, WR27, WR31 and WR32). As can be seen in Table 11.4, there are significant differences between the ratings of the 27 wine routes, ranging from a maximum of 4.862 (WR16), to a minimum of 4.373 (WR13). These differences are statistically significant ($F = 1.59$; $p < 0.05$). An analysis of the quantitative and sentiment ratings of the reviews followed the pattern identified above, that is, the ratings

Table 11.2 Quantitative and sentiment-based evaluations of wine tourism activities

	Quantitative rating (Overall)	Quantitative rating (Spanish)	Quantitative rating (English)	Sentiment-based rating (Spanish)	Sentiment-based rating (English)
Total					
Mean	4.631	4.626	4.705	5.883	10.917
SD	0.429	0.434	0.514	3.371	5.139
Max	5	5	5	35	37
Min	1	1	1	−2	−4
Wine route (WR) member					
Mean	4.642	4.654	4.703	6.191	10.787
SD	0.377	0.3814	0.509	3.215	4.561
Non-WR member					
Mean	4.622	4.605	4.706	5.652	11.019
SD	0.464	0.469	0.518	3.469	5.557

Source The authors

Table 11.3 Quantitative average ratings based on type of visit and language of the review

Type of visit	Quantitative rating (Overall)	Quantitative rating (Spanish)	*N* reviews	Quantitative rating (English)	*N* reviews
Solo	4.476	4.480	435	4.471	323
Friends	4.541	4.521	10,269	4.616	2728
Family	4.505	4.485	5330	4.590	1242
Couple	4.546	4.539	13,583	4.565	5356
Business	4.551	4.503	588	4.688	202
Not stated	4.481	4.488	5443	4.455	1845

Source The authors

given were significantly higher, for all routes, for reviews in English than for reviews in Spanish. In addition, there are differences between the best and worst rated routes based on the language of the review. Indeed, the best route for the Spanish reviews was WR5 (by quantitative and sentiment-based ratings). However, for the reviews in English, WR4, WR6, WR8 and WR24 obtained the best overall quantitative ratings, while WR24 obtained the best sentiment-based rating.

Finally, the wine routes were analysed based on visit type (see Table 11.5). The results obtained show, once again, significant differences between the ratings of the routes based on visit type. Thus, for example, WR30 is the best wine route to visit alone, while WR11 is the worst. At the destination level, this analysis makes it possible to identify some of the strengths and weaknesses of the routes based on visit type and main target market segment.

11.5 Concluding Remarks

Wine tourism is a business opportunity for wineries that can aid the economic development of the areas in which they are located. Thus, in recent years, public and private initiatives have been launched to promote wine tourism in rural areas. In Spain, wine routes play an essential role in the marketing and promotion of wine tourism. These routes provide

Table 11.4 Average of the quantitative and sentiment ratings of the wineries belonging to the different wine routes (overall, and based on the language of the review)

WR	Quantitative rating (Overall)	Quantitative rating (Spanish)	Quantitative rating (English)	Sentiment-based rating (Spanish)	Sentiment-based rating (English)
1	4.626	4.533	4.618	7.279	10.147
2	4.800	4.875	n.a	7.900	n.a
4	4.850	4.562	5.000	5.704	9.750
5	4.776	5.000	4.917	15.571	12.778
6	4.644	4.407	5.000	6.198	22.000
7	4.380	4.586	2.333	5.373	4.333
8	4.487	4.700	5.000	4.811	15.364
10	4.817	4.773	4.909	6.931	12.123
11	4.558	4.581	4.665	4.091	5.838
12	4.629	4.519	4.639	6.978	3.302
13	4.373	4.268	4.678	6.069	5.936
14	4.500	4.653	4.595	5.192	11.203
15	4.415	4.680	4.405	5.128	8.651
16	4.862	4.786	4.793	5.985	11.478
17	4.550	4.803	4.821	6.961	9.255
18	4.758	4.674	4.853	6.182	13.114
19	4.527	4.660	4.605	5.823	9.652
20	4.666	4.684	4.692	6.186	10.713
21	4.453	4.695	4.374	6.113	4.707
22	4.677	4.748	4.700	6.543	11.713
23	4.743	4.658	4.848	6.305	11.047
24	4.803	4.741	5.000	5.200	14.000
25	4.746	4.713	4.778	6.220	13.227
26	4.625	4.510	4.797	7.566	13.455

WR	Quantitative rating (Overall)	Quantitative rating (Spanish)	Quantitative rating (English)	Sentiment-based rating (Spanish)	Sentiment-based rating (English)
28	4.604	4.589	4.270	5.879	8.281
29	4.251	4.719	4.702	5.219	6.429
30	4.581	4.366	4.576	5.480	8.264

WR3, WR9, WR27, WR31 and WR32 are excluded
Source The authors

Table 11.5 Best and worst rated routes based on visit type

Type of visit	Best WRs	Worst WRs
Solo	WR30	WR11
Friends	WR8	WR9
Family	WR30/WR25	WR2
Couple	WR8	WR2
Business	WR29	WR2

Note Only the results of the scores of the Spanish languages reviews are presented
Source The authors

vehicles for the development of single collective brands and bring consistency to, and bring together, the wine tourism offer of the wineries located in wine-growing areas. With the aim of adding to the knowledge of wine tourism, this study assesses the potential impact of these wine routes on the perceptions' of wine tourists, by analysing their reviews of their visits, taking into account the influence of visit type and the nationality (Spanish/international) of the visitor (based on the language of the review).

The results showed differences in tourists' perceptions of wine tourism activities based on the route to which the winery belonged, the language of the review, and on visit type. These results have managerial implications in three areas.

First, the results have implications for how wine routes/wineries project themselves in terms of the wine tourism product itself. Although wineries have not traditionally differentiated their activities based on visitor type, as they generally have offered a single type of visit, at different times of the day, an increasing number of wineries are providing adapted offers targeted at wine tourists' profiles. The classic segmentation tool in the sector is price (different visit forms at different prices, based on service/product offering), but wineries should strive to offer activities adapted to the profiles and specific needs of their visitors. In this sense, the results of this study identify important differences in visitors' evaluations of wineries based on visit type. The lower level of satisfaction expressed by solo visitors is particularly noteworthy; it undoubtedly reflects the high social component that winery visits hold for other tourist profiles (e.g. those who visit with friends or family). This result has important implications for wineries, which should consider adapting part of their wine tourism offer to address solo visitors.

Second, the results showed that international tourists (reviews in English) were more satisfied than domestic visitors (reviews in Spanish). It should be remembered that the Spanish tourism sector depends to a large extent on international tourists, and the results of this study show that wine tourism can be a very attractive complementary product for this type of tourist and, therefore, can complement Spain's traditional sun and sand tourism product. In fact, some studies have highlighted the role of these international tourists, who traditionally buy more wine (than do Spanish tourists) during their visits to wineries. Improving their satisfaction is, therefore, key to the profitability of wine tourism.

Finally, this work has implications for wine route managers. In addition to marketing and promoting wine tourism activities, wine routes should also seek to communicate the quality of the offer. In this sense, the results of the study identified differences in tourists' perceptions of the individual wineries associated with the different wine routes, although no evidence was found that differences existed between tourists' evaluations of wineries associated with a route, and those not associated with a route. The results of the study can be used to develop benchmarking processes to identify the best rated routes, which can aid in the analysis of the most important factors/dimensions of the wine tourist's experience.

This work has limitations that should be addressed in the future. First, although TripAdvisor is one of the most popular tourism-focused review platforms, other increasingly popular review platforms (e.g. Google My Business) have very attractive functionalities that might attract potential wine tourists, so it is not possible to extrapolate our results to encompass all wine tourists' opinions. Moreover, the sentiment analysis was carried out using a specific lexicon (AFINN) which, although it has been widely used in the academic literature, is not the only one available, nor is it specific to the wine sector. Finally, the assumption was made that the reviews written in English were posted by international visitors, and that the reviews written in Spanish were posted by national visitors, although in the latter case they may have been posted by visitors from Spanish-speaking countries, for example, from Latin America.

These limitations open up new research opportunities that should be addressed in the future. Furthermore, as a promising line of research, we suggest that future studies should identify the latent dimensions of the wine tourism experience to help establish which elements of the experience have the greatest impact on overall visitor satisfaction. This evaluation of the wine tourism experience should examine the individual elements, touring the vineyard, tasting the wine, that make up the overall experience, which would allow researchers to extract specific information about the aspects of visits that tourists value. Some aspects (e.g. the wine tasting and interactions with staff) might be rated positively, while others (e.g. the appearance/cleanliness of facilities) might be rated negatively. Thus, a sentiment analysis based on specific aspects of visits could provide a broader and more accurate picture of how wine tourists assess the activity.

Funding This work was partially supported by the Spanish Ministry of Science and Innovation under research project SOSTUR (PID2022-141694NB-I00).

References

Alonso, A. D., Fraser, R. A., & Cohen, D. A. (2007). Does age matter? How age influences the winery experience. *International Journal of Culture, Tourism and Hospitality Research, 1*(2), 131–139. https://doi.org/10.1108/17506180710751678

Alonso, A. D., & Liu, Y. (2012). Coping with changes in a sector in crisis: The case of small Spanish wineries. *Journal of Wine Research, 23*(1), 81–95. https://doi.org/10.1080/09571264.2011.646252

Angelini, A., & Gilli, A. (2021). Customer experience can play a strategic role for wineries. *Journal of Business Strategy*. https://doi.org/10.1108/JBS-06-2021-0103

Barbierato, E., Bernetti, I., & Capecchi, I. (2021). Analyzing TripAdvisor reviews of wine tours: An approach based on text mining and sentiment analysis. *International Journal of Wine Business Research*. https://doi.org/10.1108/IJWBR-04-2021-0025

Boo, S., Busser, J., & Baloglu, S. (2009). A model of customer-based brand equity and its application to multiple destinations. *Tourism Management, 30*(2), 219–231. https://doi.org/10.1016/j.tourman.2008.06.003

Bruwer, J. (2003). South African wine routes: Some perspectives on the wine tourism industry's structural dimensions and wine tourism product. *Tourism Management, 24*(4), 423–435. https://doi.org/10.1016/S0261-5177(02)00105-X

Byrd, E. T., Canziani, B., (Jerrie) Hsieh, Y.-C., Debbage, K., & Sonmez, S. (2016). Wine tourism: Motivating visitors through core and supplementary services. *Tourism Management, 52*, 19–29. Scopus. https://doi.org/10.1016/j.tourman.2015.06.009

Caldwell, N., & Freire, J. R. (2004). The differences between branding a country, a region and a city: Applying the brand box model. *Journal of Brand Management, 12*(1), 50–61. https://doi.org/10.1057/palgrave.bm.2540201

Cardinale, S., Nguyen, B., & Melewar, T. (2016). Place-based brand experience, place attachment and loyalty. *Marketing Intelligence & Planning, 34*(3). https://doi.org/10.1108/MIP-04-2014-0071

Carlsen, J. (2004). A review of global wine tourism research. *Journal of Wine Research, 15*(1), 5–13. https://doi.org/10.1080/0957126042000300281

Carlsen, J., & Boksberger, P. (2015). Enhancing consumer value in wine tourism. *Journal of Hospitality & Tourism Research, 39*(1), 132–144. https://doi.org/10.1177/1096348012471379

Charters, S., & Ali-Knight, J. (2002). Who is the wine tourist? *Tourism Management, 23*(3), 311–319. https://doi.org/10.1016/S0261-5177(01)00079-6

Charters, S., & Fountain, J. (2006). Younger wine tourists: A study of generational differences in the cellar door experience. In J. Carlsen & S. Charters (Eds.), *Global wine tourism: Research, management and marketing* (pp. 153–160). CABI. https://doi.org/10.1079/9781845931704.0153

Dawson, H., Holmes, M., Jacobs, H., & Wade, R. I. (2011). Wine tourism: Winery visitation in the wine appellations of Ontario. *Journal of Vacation Marketing, 17*(3), 237–246. https://doi.org/10.1177/1356766711409185

Dodd, T. H. (1995). Opportunities and pitfalls of tourism in a developing wine industry. *International Journal of Wine Marketing, 7*(1), 5–16. https://doi.org/10.1108/eb008636

Famularo, B., Bruwer, J., & li, E. (2010). Region of origin as choice factor: Wine knowledge and wine tourism involvement influence. *International Journal of Wine Business Research, 22*(4), 362–385. Scopus. https://doi.org/10.1108/17511061011092410

Fotiadis, A., Polyzos, S., & Huan, T.-C.T.C. (2021). The good, the bad and the ugly on COVID-19 tourism recovery. *Annals of Tourism Research, 87*, 103117. https://doi.org/10.1016/j.annals.2020.103117

Gao, S., Tang, O., Wang, H., & Yin, P. (2018). Identifying competitors through comparative relation mining of online reviews in the restaurant industry. *International Journal of Hospitality Management, 71*, 19–32.

Gatti, S., & Incerti, F. (1997). *The wine routes as an instrument for the valorisation of typical products and rural areas* (52nd Seminar, June 19–21, 1997, Parma, Italy N.° 231431). European Association of Agricultural Economists. https://econpapers.repec.org/paper/agseaae52/231431.htm

Getz, D. (2000). *Explore wine tourism: Management, development and destinations.* Cognizant Communication.

Getz, D., & Brown, G. (2006). Critical success factors for wine tourism regions: A demand analysis. *Tourism Management, 27*(1), 146–158. https://doi.org/10.1016/j.tourman.2004.08.002

Gómez, M., González-Díaz, B., & Molina, A. (2015a). Priority maps at wine tourism destinations: An empirical approach in five Spanish wine regions. *Journal of Destination Marketing & Management, 4*(4), 258–267. https://doi.org/10.1016/j.jdmm.2015.09.003

Gómez, M., Lopez, C., & Molina, A. (2015b). A model of tourism destination brand equity: The case of wine tourism destinations in Spain. *Tourism Management, 51*, 210–222. https://doi.org/10.1016/j.tourman.2015.05.019

Gómez, M., Pratt, M. A., & Molina, A. (2019). Wine tourism research: A systematic review of 20 vintages from 1995 to 2014. *Current Issues in Tourism, 22*(18), 2211–2249. https://doi.org/10.1080/13683500.2018.1441267

Gross, M. J., & Brown, G. (2008). An empirical structural model of tourists and places: Progressing involvement and place attachment into tourism. *Tourism Management, 29*(6), 1141–1151. https://doi.org/10.1016/j.tourman.2008.02.009

Hall, C. M. (2002). *Wine tourism around the world: Development, management and markets.* Elsevier.

Hall, C. M., & Macionis, N. (1998). Wine tourism in Australia and New Zealand. In R. W. Butler, C. M. Hall, & J. M. Jenkins (Eds.), *Tourism and recreation in rural areas* (pp. 267–298). Wiley.

Hall, C., & Mitchell, R. (2002). The tourist terroir of New Zealand wine: The importance of region in the wine tourism experience. *Food and Environment: Geographies of Taste*, 69–91.

Hojman, D. E., & Hunter-Jones, P. (2012). Wine tourism: Chilean wine regions and routes. *Journal of Business Research, 65*(1), 13–21. https://doi.org/10.1016/j.jbusres.2011.07.009

Konecnik, M., & Gartner, W. C. (2007). Customer-based brand equity for a destination. *Annals of Tourism Research, 34*(2), 400–421. https://doi.org/10.1016/j.annals.2006.10.005

Leri, I., & Theodoridis, P. (2021). How do personality traits affect visitor's experience, emotional stimulation and behaviour? The case of wine tourism. *Tourism Review, 76*(5), 1013–1049. https://doi.org/10.1108/TR-05-2019-0148

López-Guzmán, T., Vieira-Rodríguez, A., & Rodríguez-García, J. (2014). Profile and motivations of European tourists on the Sherry wine route of Spain. *Tourism Management Perspectives, 11*, 63–68. https://doi.org/10.1016/j.tmp.2014.04.003

Marzo-Navarro, M., & Pedraja-Iglesias, M. (2010). Are there different profiles of wine tourists? An initial approach. *International Journal of Wine Business Research, 22*(4), 349–361. https://doi.org/10.1108/17511061011092401

Mitchell, R., & Hall, C. M. (2006). Wine tourism research: The state of play. *Tourism Review International, 9*(4), 307–332. https://doi.org/10.3727/154427206776330535

WRA. (2021). *Jornada de enoturismo.* Ruta del Vino de Alicante, 28 de octubre. Universidad de Alicante.

Sellers, R., & Nicolau, J. L. (2021). Satisfaction and expenditure in wineries: A prospect theory approach. *Journal of Hospitality & Tourism Research, 47.*https://doi.org/10.1177/10963480211031407

Sparks, B. (2007). Planning a wine tourism vacation? Factors that help to predict tourist behavioural intentions. *Tourism Management, 28*(5), 1180–1192. https://doi.org/10.1016/j.tourman.2006.11.003

Tahar, Y. B., Haller, C., & Massa, C. (2021). Business tourism in the wine sector: An exploratory study. *Journal of Wine Research, 32*(4), 262–280. https://doi.org/10.1080/09571264.2021.2012442

Telfer, D. J. (2001). Strategic alliances along the Niagara wine route. *Tourism Management, 22*(1), 21–30. https://doi.org/10.1016/S0261-5177(00)00033-9

Thomas, B., Quintal, V. A., & Phau, I. (2018). Wine tourist engagement with the winescape: Scale development and validation. *Journal of Hospitality & Tourism Research, 42*(5), 793–828. https://doi.org/10.1177/1096348016640583

Veres, D., Clark, H., & Golbourne, D. (2008). Increasing the contribution of special events to Niagara's tourism industry. *International Journal of Contemporary Hospitality Management, 20*(3), 313–318. https://doi.org/10.1108/09596110810866127

Vo Thanh, T., & Kirova, V. (2018). Wine tourism experience: A netnography study. *Journal of Business Research, 83*, 30–37. https://doi.org/10.1016/j.jbusres.2017.10.008

Williams, P. W., & Dossa, K. B. (2003). Non-resident wine tourist markets. *Journal of Travel & Tourism Marketing, 14*(3–4), 1–34. https://doi.org/10.1300/J073v14n03_01

Williams, P. W., & Kelly, J. (2001). Cultural wine tourists: Product development considerations for British Columbia's resident wine tourism market. *International Journal of Wine Marketing, 13*(3), 59–76. https://doi.org/10.1108/eb008727

Index

J. Martínez-Falcó et al. (eds.), *Strategic Management in the Wine Tourism Industry*, https://doi.org/10.1007/978-3-031-54837-6

GPSR Compliance

The European Union's (EU) General Product Safety Regulation (GPSR) is a set of rules that requires consumer products to be safe and our obligations to ensure this.

If you have any concerns about our products, you can contact us on ProductSafety@springernature.com

In case Publisher is established outside the EU, the EU authorized representative is:

Springer Nature Customer Service Center GmbH
Europaplatz 3
69115 Heidelberg, Germany

The manufacturer's authorised representative in the EU is Springer
Nature Customer Service Centre GmbH, Europaplatz 3, 69115 Heidelberg,
Germany. If you have any concerns regarding our products, please
contact ProductSafety@springernature.com

Printed and bound by CPI Group (UK) Ltd, Croydon, CR0 4YY
01/10/2026
02225898-0004